D1760197

Climate Change and the Crisis of Capitalism

Are established economic, social and political practices capable of dealing with the combined crises of climate change and the global economic system? Will falling back on the wisdoms that contributed to the crisis help us to find ways forward or simply reconfigure risk in another guise? This volume argues that the combination of global environmental change and global economic restructuring require a rethinking of the priorities, processes and underlying values that shape contemporary development aspirations and policy.

This volume brings together leading scholars to address these questions from several disciplinary perspectives: environmental sociology, human geography, international development, systems thinking, political sciences, philosophy, economics and policy/management science. The book is divided into four parts that examine contemporary development discourses and practices. It bridges geographical and disciplinary divides, and includes chapters on innovative governance that confront unsustainable economic and environmental relations in both developing and developed contexts. It emphasises the ways in which dominant development paths have necessarily forced a separation of individuals from nature, but also from society and even from 'self'. These three levels of alienation each form a thread that runs through the book. There are different levels and opportunities for a transition towards resilience, raising questions surrounding identity, governance and ecological management. This places resilience at the heart of the contemporary crisis of capitalism, and speaks to the relationship between the increasingly global forms of economic development and the difficulties in framing solutions to the environmental problems that carbon-based development brings in its wake. Existing social science can help in not only identifying the challenges but also potential pathways for making change locally and in wider political, economic and cultural systems, but it must do so by identifying transitions out of carbon dependency and the kinds of political challenges they imply for reflexive individuals and alternative community approaches to human security and well-being.

Climate Change and the Crisis of Capitalism contains contributions from leading scholars to produce a rich and cohesive set of arguments, from a range of theoretical and empirical viewpoints. It analyses the problem of resilience under existing circumstances, but also goes beyond this to seek ways in which resilience can provide a better pathway and template for a more sustainable future. This volume will be of interest to both undergraduate and postgraduate students studying Human Geography, Environmental Policy and Politics.

Mark Pelling is a Professor of Geography at King's College, London. His research specialism is adaptation to climate change, in developing countries and more recently in the UK and Europe. **David Manuel-Navarrete** is a Senior Research Associate at King's College, London and a Visiting Researcher at desiguALdade.net (Free University of Berlin and Ibero-American Institute), where he studies spatial inequalities created by global tourism in the Mexican Caribbean. **Michael Redclift** is Professor of International Environmental Policy at King's College, London. His research interests include sustainable development, global environmental change, environmental security and the modern food system.

Routledge studies in human geography

This series provides a forum for innovative, vibrant and critical debate within Human Geography. Titles will reflect the wealth of research which is taking place in this diverse and ever-expanding field.

Contributions will be drawn from the main subdisciplines and from innovative areas of work which have no particular subdisciplinary allegiances.

Climate Change and the Crisis of Capitalism

A chance to reclaim self, society and nature

**Edited by Mark Pelling,
David Manuel-Navarrete and
Michael Redclift**

 Routledge
Taylor & Francis Group

LONDON AND NEW YORK

First published 2012
by Routledge
2 Park Square, Milton Park, Abingdon, Oxon OX14 4RN

Simultaneously published in the USA and Canada
by Routledge
711 Third Avenue, New York, NY 10017

Routledge is an imprint of the Taylor & Francis Group, an informa business

© 2012 Selection and editorial matter: Mark Pelling, David Manuel-
Navarrete and Michael Redclift; individual chapters, the contributors

The right of Mark Pelling, David Manuel-Navarrete and Michael Redclift
to be identified as the authors of the editorial material, and of the authors
for their individual chapters, has been asserted in accordance with sections
77 and 78 of the Copyright, Designs and Patents Act 1988.

All rights reserved. No part of this book may be reprinted or reproduced or
utilised in any form or by any electronic, mechanical, or other means, now
known or hereafter invented, including photocopying and recording, or in
any information storage or retrieval system, without permission in writing
from the publishers.

Trademark notice: Product or corporate names may be trademarks or
registered trademarks, and are used only for identification and explanation
without intent to infringe.

British Library Cataloguing in Publication Data
A catalogue record for this book is available from the British Library

Library of Congress Cataloging in Publication Data
Climate change and the crisis of capitalism/edited by Mark Pelling, David
Manuel-Navarrete and Michael Redclift.
   p. cm.
   Includes bibliographical references and index.
   1. Climatic changes–Economic aspects. 2. Climatic changes–Political
aspects. 3. Capitalism. 4. Global environmental change. I. Pelling,
Mark. II. Manuel-Navarrete, David. III. Redclift, M. R.
   QC902.9.C55 2011
   363.738'74–dc23

ISBN: 978-0-415-67694-6 (hbk)
ISBN: 978-0-203-14611-8 (ebk)

Typeset in Times New Roman
by Wearset Ltd, Boldon, Tyne and Wear

BLACKBURN COLLEGE
LIBRARY
Acc. No. BB52461
Class No. *UCL 363.7387PEL*
Date *01-11-12*

Contents

Contributors

Ian Bailey is Associate Professor in Human Geography at the University of Plymouth, specialising in UK and European Union climate policy and politics. He has published one authored volume, *New Environmental Policy Instruments in the European Union* (Ashgate), a special issue of *Area* on climate policy implementation, and co-edited *Turning Down the Heat: The Politics of Climate Policy in Affluent Democracies*, with Hugh Compston (Palgrave Macmillan, 2008). He has advised UK government, EU committees, the World Bank and Policy Network on various aspects of climate policy.

John Barry is Reader in Politics and Associate Director of the Institute for a Sustainable World at Queens University Belfast. His publications include *Rethinking Green Politics* (Sage, 1999) [Winner of the PSA's WJM Mackenzie Prize for best book published in political science 1999] and *Environment and Social Theory*, 2nd edn (Routledge, 2007).

Katrina Brown is Professor of Development Studies at the University of East Anglia. She is part of the Tyndall Centre for Climate Change Research and led the International Development Programme. She is co-editor of the journal *Global Environmental Change*. She currently holds an ESRC Professional Fellowship on Resilient Development in Social Ecological Systems (2009–2012) which aims to advance theoretical and conceptual understanding of resilience across the social sciences.

Marcus Carson is an Associate Professor at Stockholm University and a Senior Research Fellow at the Stockholm Environment Institute. His research focuses on the sociology of public policy, environmental and political sociology, neo-institutional theory, policy paradigms and climate change policy, especially in the US and EU. He is co-author of *Paradigms in Public Policy*.

Hugh Compston is Professor of Politics at Cardiff University and has published widely on political economy, the future of public policy and climate politics. Since 2007 he has led an international academic investigation into the politics of climate change. Results so far include *Turning Down the Heat: The*

Politics of Climate Policy in Affluent Democracies, co-edited with Ian Bailey (Palgrave Macmillan), and a special book issue of *Environmental Politics* entitled 'Climate Policy and Political Strategy'.

Andy Gouldson is Director of the Centre for Climate Change Economics and Policy at the University of Leeds. Having worked extensively on ecological modernisation and on the influence of different forms of environmental policy and governance, his work now focuses on the prospects for a low carbon economy and society, with a particular emphasis on the prospects for and limits to decarbonisation through the fine tuning of existing systems and institutions.

Mattias Hjerpe is Assistant Professor at the Centre for Climate Science and Policy Research and Department of Water and Environmental Studies, Linköping University. He focuses on the interaction of climate and socioeconomic stressors, climate and sustainable development policy linkages, climate change governance and utopian thought.

Björn-Ola Linnér is Professor in Water and Environmental Studies and at the Centre for Climate Science and Policy Research, Linköping University. He focuses on integration of climate and sustainable development policies, food security, transnational governance and utopian thought. Publications include *The Return of Malthus: Environmentalism and Postwar Population–Resource Crises* (2003).

David Manuel-Navarrete is a Visiting Researcher at desiguALdade.net (Free University of Berlin and Ibero-American Institute), where he studies the spatial inequalities created by global tourism in the Mexican Caribbean. He holds degrees in environmental sciences, ecological economics (Autonomous University of Barcelona) and geography (University of Waterloo, Canada). He has collaborated with the universities of Chiang Mai, Harvard, Stanford, Arizona State and Calgary through international research projects. Between 2004 and 2007, He worked at the United Nations Economic Commission for Latin America and the Caribbean (ECLAC). He is lead author on over a dozen journal and book chapter publications.

Diana Mitlin works at the International Institute for Environment and Development and the Institute for Development Policy and Management (University of Manchester). Her work focuses on urban poverty and inequality, including urban poverty reduction programmes and the contribution of collective action by low-income and otherwise disadvantaged groups. Publications include *Empowering Squatter Citizen* (Earthscan, 2004, with David Satterthwaite), *Confronting the Crisis in Urban Poverty* (ITDG, 2006, with Lucy Stevens and Stuart Coupe), *Can NGOs make a Difference? The Challenge of Development Alternatives*, (Zed Books, 2007, with Tony Bebbington and Sam Hickey), and *Rights-based Approaches to Development: The Pitfalls and Potentials* (Kumarian Press, 2008, with Sam Hickey).

Peter North is Senior Lecturer in Geography at the University of Liverpool. He is a long-time activist in and researcher of local economic trading systems and other forms of alternative economic and trading systems in the UK and internationally. He is a member of the editorial board of the *International Journal of Complementary Currency Research*. Recent publications include *Money and Liberation: The Micropolitics of Alternative Currency Movements* (University of Minnesota Press, 2007) and *Alternative Currencies as a Challenge to Globalisation? A Case Study of Manchesters Local Currency Networks* (Ashgate, 2005).

Mark Pelling is Professor of Geography at King's College London. His research specialism is adaptation to climate change, in developing countries and more recently in the UK and Europe. He is a lead author for the IPCC AR5 and the IPCC SREX Report. His publications include *Adaptation to Climate Change: From Resilience to Transformation* (Routledge, 2010) and *The Vulnerability of Cities: Social Resilience and Natural Disaster* (Earthscan, 2003). He has consulted on adaptation and disaster risk reduction issues for several agencies including the UK Environment Agency, DFID, UNDP and UN-HABITAT.

Michael Redclift is Professor of International Environmental Policy, Department of Geography, King's College London. Research interests include sustainable development, global environmental change, environmental security and the modern food system. In 2006 he was the first recipient of the 'Frederick Buttel Award', from the International Sociological Association. Recent books include Editor, *Sustainability: Critical Concepts*, (four volumes) (Routledge, 2005), *Frontiers: Histories of Civil Societies and Nature* (MIT Press, 2006), and *Climate Change and Human Security* (Edward Elgar, 2011).

Molly Scott Cato is Reader in Green Economics, Cardiff Metropolitan University. She specialises in the issues of trade, work, money and cooperatives, and is interested in bioregionalism as an economic response to climate change. She is a member of the Executive of the UK Society of Cooperative Studies and, from 2007, Director of the Cardiff Institute for Cooperative Studies. Recent publications include *Environment and Economy* (Routledge, 2011) and *Green Economics: Theory, Policy and Practice* (Earthscan, 2009), Molly is a Director of Transition Stroud, helped to establish the Stroud Pound, and works as an engaged green economist in her local community of Stroud, Gloucestershire.

Rory Sullivan is a Senior Research Fellow at the University of Leeds and, together with Professor Andy Gouldson, is leading the ESRC Centre for Climate Change Economics and Policy project Non-State Actors and the Low Carbon Economy. Rory is an internationally recognised expert on responsible investment and climate change and holds a number of advisory roles on these issues. He has written numerous books, papers and articles on investment, climate change, human rights and development issues. His books include the newly published *Valuing Corporate Responsibility: How Do Investors Really Use Corporate Responsibility Information?* (Greenleaf Publishing, 2011) and the edited collection *Responsible Investment* (Greenleaf Publishing, 2006).

1 Climate change and the crisis of capitalism

Mark Pelling, David Manuel-Navarrete and Michael Redclift

Introduction

Are established economic, social and political practices capable of dealing with the combined contemporary crises of climate change and intensifying economic inequality and global economic disruption? Will falling back on those wisdoms that have prefigured individual crises help identify ways forward, or simply reconfigure risk so that it may reappear in another guise in the future? This volume argues that the combination of global environmental change and the global economic downturn provides an opportunity for critical thinking and policy formulation by highlighting the co-dependence of socio-political and ecological processes. Crisis in this understanding signifies a point of instability in predominant structures, a precursor to impending threat, but importantly also an opportunity for the consideration and emergence of alternatives. Our starting point is a concern for global environmental change, which remains in the foreground of analysis, but we argue cannot be understood without engaging also with the drivers and consequences of current economic rounds of restructuring.

A critical research agenda on crisis and risk management is called for, one that can approach risk management through development agendas, not as a stand-alone policy archipelago (Hewitt, 1983). Some progress has been made in examining the interdependencies of environmental risk and human development. Beck (1992, 2008) famously describes the rise of risk as a driving force for decision-making in Western consciousness and beyond as the Risk Society. One defining quality of risk society is the disassociation of risk and everyday life as hazards become harder to detect without scientific techniques and so more difficult to connect with existing popular and political movements and agendas for change or resistance (Beck *et al.*, 1994). Global environmental change and climate change in particular are certainly examples of this with the scale as well as temporal dimensions of both making them invisible to daily life yet simultaneously, and increasingly formative of it and configured by it.

This is a challenge of alienation and separation. In effect, there is an existential gap between what can be done to confront economic and environmental challenges (and there is much that could be done), and what culture and power determine is reasonable and proper for society to do. Some have likened to addiction society's failure to acknowledge, let alone act as individuals or collectively, to address the combined economic and environmental crisis (Eckersley, 2007). This may be a pertinent metaphor but an over-simplistic one as well. The current crisis is so embedded in everyday behaviour justified by co-produced values and reinforced through habit that the very constructions of identity and notions of self – the signifiers of success, happiness, status, to say nothing of norms of social responsibility – have become part of the problem. This is demonstrably so in affluent societies displaying excess consumption but also in poorer societies where the direction of aspiration is clear.

In responding to this challenge, one that transcends the usual disciplinary boundaries of politics, economics, sociology or psychology, we present an agenda that questions contemporary power relations and opens up the normative underpinnings of dominant and alternative modes of accumulation and social reproduction. This responds to a gap in predominant social science work on global environmental change which has, thus far, tended to focus on the detail of problem-solving or on mechanisms for changing individual attitudes and behaviour. This approach is in danger of missing the bigger picture – of describing capacities for incremental reform rather than processes through which alternative histories have and may unfold. A second trend in established thinking has been to present defensive visions of responding to change – be this in terms of resilience or adaptation – that seek perversely to protect the status quo. These conceptual contradictions are underlain by limitations in existing analytical models that are inadequate, for example, in being unable to fully (or in some cases to even partially) capture power, competing visions and viewpoints, and inner worlds of emotion in analysing crisis and our relationships to it.

This book asks how far the continued application of terms like resilience can improve our conceptualisation and empirical observation of socio-political structures and agency during crises – when such crises are themselves embedded in ongoing unsustainable and unjust development trajectories. Ultimately, we are interested in the prospects for new conceptualisations in framing transformation and alternative approaches to policy and political action, which promote ecological integrity, and procedural and distributional equity as part of living with and beyond crisis.

This volume brings together 14 contributors, each a well-known expert and presenting a specific viewpoint on our common challenge. This range of viewpoints is necessary, as this introductory chapter argues, not only because the current crisis is multifaceted but also because it is part of the evolving history of multiple strands in human and environmental sciences. Getting to grips with the nature of crisis and thinking through responses that can enhance social justice and environmental security therefore requires a multidisciplinary envisioning.

In addition to describing the character and drivers for the current expression of the continuing ecological-economic crisis we aim to explore the scope for pathways that can offer escape routes from ever more perpetual rounds of crisis and response. This includes analysis of existing policy frameworks and political decision-making at levels from the international to the individual, including examples of bottom-up, self-liberation. In all cases we find culture and identity are as important as economy and politics as resources and motors for change (and also of course as barriers to change). An approach that can examine the co-evolution of capitalism from multiple standpoints is also an essential first step in recognising the equally varied and dynamic qualities of any potential responses that can confront the internal contradictions of capitalism (its drive for creative destruction), question core assumptions (growth) and identify alternatives in everyday life as well as those open to political and economic policy or market forces (from liberating social movements to ecological modernisation and the 'new green deal'). To achieve this, the book integrates perspectives from many viewpoints: development studies, socio-ecological systems theory, political theory, international relations, risk management and environmental sociology. Following this introductory chapter the book is organised into four parts. Part I reviews the conceptual and policy constraints imposed by dominant interpretations of sustainable development and resilience. Part II examines the potential of already existing alternatives that challenge established hierarchies in the power-knowledge nexus through localised and everyday actions. Parts III and IV deploy organisational, economic and political theory to explore the potential of macro-economic and political alternatives and scope for a new politics of climate change.

Within this chapter we frame the subsequent sections by first revisiting the social construction of global environmental change and then presenting three elements that make up a broad meta-framework within which to position the more detailed following chapters. The framework is derived from co-evolutionary theory, discourses of austerity and fear, and a revision of socio-ecological agency and alienation.

What is global environmental change?

Global environmental change is but one facet of capitalism's current crisis; a crisis we would do well to examine in its broader manifestation to help reveal the cause-and-symptom positionality of social and environmental change. Beforehand it is worth noting that contemporary capitalism is not alone in having co-produced social and environmental degradation to the point of crisis. In the twentieth century, state socialism as practised under the Soviet Union shared with Western capitalism a modernist agenda driven by industrialisation that equally well alienated nature and citizens from development, allowing ecological exploitation and social control in the name of national economic growth (Bowers, 1993). Evidence of the role of human-induced crises in agricultural societies has also shown the tendency for humanity to exploit natural assets to the point of (self)destruction. In pre-modern civilisations, such as the classic Maya of Central America, increasing rates of natural resource consumption fuelled by demographic growth and internal cultural tensions, and at times compounded by dangerous periods in natural cycles (e.g. in precipitation), catalysed collapse (Sharer, 2006). Although separated by ideology, geography and history, these systems crises demonstrate the co-dependence of natural and social life, with crisis arising in large part from a failure of institutions to adequately arbitrate relationships both within society and between society and nature. While there may be little novelty in the advent of socio-ecological tensions as a component of systems crisis, its contemporary expression is unprecedented in the scale and the complexity of its dynamics, providing significant challenges for any efforts at reform or more fundamental change.

What then are the other expressions of global capitalism's current crisis? These are well known but seldom placed alongside global environmental change. They include growing economic inequality between and within nations, the global economic downturn, and public disengagement and marginalisation from the political process. These are facets of crisis that feed into one another so that symptom and response become cause over time.

Inequality expresses itself in many ways – through access to basic services, income and consumption. Privatisation of core public services such as water provision, energy production and increasingly health care and education have provided new financial investments but these are not equally distributed, either in society or geographically. Economic globalisation has not been equal and continues to distort economic relations, with more powerful states protecting the interests of national economic lobbies. At the same time, the movable quality of investment capital has contributed to a transfer of power from states to capital as states compete with varying success for international capital investment (Dicken, 2007).

The financial sector crisis that triggered the current global economic downturn had its roots in responses to previous cycles of crisis and growth (Harvey, 2010). A driving belief in the need for the global economy to grow annually by 3 per cent produces a constant need for new areas of investment and return. A focus on stock capital including speculation over property value overheated by debt is but the most recent example. Previous crises have been offset by investment opportunities in technological innovation (computing and the internet) and reconstruction following conflict. Now, with public fiscal debt burdens causing cuts in public sector service provision and public sector wage and pension settlements, economic downturn is compounding inequality further.

One might suppose that inequality and economic uncertainty would galvanise popular political movements worldwide. This has rarely been the case. An exception was Spanish popular protest in spring 2011. A response to deep public sector retrenchment proposals in a context of high youth unemployment, and perhaps inspired by the Arab spring uprisings in North Africa, these protests succeeded in politicising a generation of otherwise alienated Spanish youth. More generally, however, political disengagement continues to be the rising norm in democratic, Western societies (Chatterjee, 2007). This is explained variously as a result of the complexity of evidence-based policy (e.g. climate change), the rise of celebrity over policy in popular media, or the shift to middle-ground politics where there is little beyond managerial style to separate political parties. Although some (e.g., Giddens) see the drift to the middle ground as an advantage in offering hope for cross-party consensus in responding to climate change (Giddens, 2009). In part the democratic deficit in Western states and elsewhere is indicative of the public's recognition of the movement of power from government to intergovernmental organisations and transnational private capital. This was seen most recently in the failure of US regulators to properly review deep-water-drilling safety procedures in BP's Gulf of Mexico operations. The reluctance of government to regulate an increasingly technically complex industry and one located at the core of the state–capital nexus has distinct echoes of the failures of regulation lying in part behind the global banking collapse of 2008/2009. Elsewhere, repressive political regimes continue to maintain tight control over political dissent, for example, in China, Myanmar and Thailand. These models of authoritarian state capitalism are more direct and forceful but no less alienating than their Western market capitalism counterparts. There is scope for learning from organised resistance in both political variants.

Global environmental change itself is a function of consumption and demography mediated by local development choices and technological innovation. This includes changes taking place at the global scale (e.g. in atmospheric conditions and ocean chemistry and loss of biodiversity) and local changes (e.g. soil loss and deforestation) that in aggregate amount to global impacts. As a basic site for the investment of wealth and its continued accumulation as well as a reservoir of ecosystem services for human health, natural assets and systems have been vastly transformed by capital (Smith, 2010). Many locally degraded systems have already crossed thresholds into collapse with the compound impact of global environmental change increasing uncertainty and risk (MEA, 2005); risk that comes from slow-changing long-term stresses as well as more frequent or unexpected extreme shocks.

These four faces of capitalism's crisis have tended to be separated in policy and academic analysis. This makes some sense – each element is complicated – but this approach makes it hard to identify the interaction and potentially reinforcing qualities of systems elements. One outcome of this reductive approach has arguably been to fall upon apparent solutions to one aspect of crisis that simply generates the foundations for a new crisis elsewhere. Like a bucket full of holes, efforts to address one expression of crisis – to plug one hole – may serve only to increase pressure elsewhere. Unlike the laws of physics, human decision-making is not value neutral; in fact, short-term and partial remedies to manifest crisis dominate because they best serve established value priorities, including the expressed values of poorer and more marginalised populations (who from positions of exposure and vulnerability are often least able and willing to face the uncertainties implicit in radical change or even incremental reform) as well as those of dominant interests. The trap of falling back on established wisdoms to confront new expressions of crisis is seductive in these socio-political contexts, and in the short-term responds to the pressures of risk aversion. Ultimately though, this strategy, while providing serial, short-term stability, runs the risk of reproducing or even strengthening those wider systemic features that are the root causes of compound crises until thresholds are breached, forcing potentially catastrophic systemic change (Pelling, 2010). To follow Naomi Klein (2007), this is a strategy for accelerating the status quo – one that generates risk in the first place.

From co-evolution to co-revolution?

The simultaneous co-production and interdependence of social, economic, political, technological and environmental aspects of history was recognised by Norgaard (1994) in his thesis on co-evolution. Here he developed co-evolution to help explain the messiness of history and in particular the failed promise of linear social progress. Co-evolution uses the metaphoric language of evolutionary biology to describe the interaction of social and socio-ecological systems elements. Norgaard's particular interest is in using this frame to explain the relative success (longevity and influence) or failure of individual development pathways that emerge out of or in response to systems elements interaction. Our current addiction to fossil fuels is presented as a co-evolutionary dead-end (Norgaard, 1995). In this reading of history, the apparent liberation from nature and time provided by carbon-intensive agricultural and transport technologies is a distortion. False liberation from nature becomes a cause of future crises as it transforms over time into alienation.

Norgaard's (1994) thesis includes knowledge and values alongside technology, social organisation and the natural environment as categories, sites and drivers for adaptation. He also successfully moves from a materialist (adaptation can be described through technical changes (e.g. in engineering or farming practices)) to a relational and constructivist epistemology, where adaptation includes changes in identity and well-being as well as humanity's relation with the non-human (Pelling, 2010). The abstract nature of co-evolution makes for difficult translation into an empirical research framework. While co-evolution has been successful at the level of metaphor to frame accounts of adaptive behaviour within complex systems (Pelling, 2003) and economic-ecological systems interaction at the global scale (Schneider and Londer, 1984), it has more limited applicability as a tool for local analysis. One useful line of analysis is the relationship between intention (policies) and emergence (self-organised activity) in policy sectors, the latter in large part accounting for observed divergence from policy during implementation (Sotarauta and Srinivas, 2006), and so revealing tensions between the actions and values of competing adaptive strategies or other behaviour.

In his analysis of contemporary capitalism, Harvey (2010) also turns to an evolutionary metaphor. History is an outcome of the interaction of seven activity spheres (technologies and organisational forms, social relations, institutional and administrative arrangements, production and labour processes, relations to nature, the reproduction of daily life and mental conceptions of the world). Harvey invests a little more independence in his components than Norgaard but both see history as determined by their dynamic interaction. Both also argue that it is in the spaces created by these dynamic interactions that the opportunities and risks of surprise are often found, and including those which have given shape to our current crisis. Nature, for example, is perpetually changing often in unexpected ways and under the influence of human intervention. Mental conceptions of the world are subject to fashion, the wax and wane of competing ideological, religious or cultural beliefs. These in turn have consequences for the acceptability of technology, labour processes and administrative arrangements influencing the social and spatial distribution of power. Geography matters here insofar as local contexts give material and temporal fixity to the activity spheres with global outcomes unfolding differently but with common roots in London, Lagos and Louisiana.

Using a co-evolutionary frame, the search for responses to crises begins with the realisation that existing policy approaches apply incomplete lenses. Solutions to the environmental crisis are seen as lying in the technological or at the margins of the economic realm. Scientific innovation and shifts in the institutions affecting economic behaviour are the tools for action. Important though they undoubtedly are, this worldview glosses over deeper tensions in organisational, social and psychological life. Co-evolution argues that no one realm of activity is dominant in history and consequently none should dominate in the casting of alternative pathways. The 1960s saw labour relations dominate, while the 1990s placed technological and organisational innovation at the heart of struggles for change. Harvey's (2010) call for a co-revolutionary agenda has this as its starting point – that the initial point of entry for alternatives is less important than the need to infect and influence other domains. Such shifts and movements are not minor historical events and most likely require energies both at the grassroots as well as momentum from above. This is a central challenge.

The structural barriers that determine the durability of a dominant system even in the face of crisis were perceptively described by Handmer and Dovers (1996), who distinguish between three cultural-political models: resistance and maintenance; change at the margins; and openness and adaptability. Resistance and maintenance are commonplace, particularly within authoritarian political contexts where access to information is controlled. It is characterised by resistance to change: actors deny risk or delay action through a call for greater scientific research before action is possible. Vulnerability can be held at bay by technical innovation, even if this generates additional risks for other places or times. This mode of crisis management generates little threat to the status quo; however, when overcome, the system is threatened with almost complete collapse. Change at the margins is perhaps the most common response to environmental threat. Risk is acknowledged and adaptations undertaken, but is limited to those that do not threaten core attributes of the dominant system as defined by the elite. Advocates argue that this form of resilience offers an incremental reform, but it may serve only to delay more major reforms by offering a false sense of security. Preference for near-term stability over radical reform for the well-being of future generations provides a strong incentive for this form of resilience. Social systems displaying openness and adaptability tackle the root causes of risk, are flexible, and are prepared to change direction rather than resist change in the face of uncertainty. That this mode of resilience is so rare is testament to the huge inertia that results from personal and collective investment in the status quo. Large fixed capital investments make change difficult as do investments in soft infrastructure – preferences for certain types of education or cultural values make shifts painful in industrial societies. Dangers also lie with instability and the likelihood of some counterproductive and unexpected outcomes. Handmer and Dovers caution that most actors will operate in only a small part of this range. This points to a central challenge for those seeking alternatives: that the space for action is relatively small because both those with power and the marginalised are wary of the instability they anticipate from significant social change.

Discourses of austerity and fear: climate change and war

It is our contention that the discussion of the crisis surrounding climate change is necessarily linked today with the causes and outcomes of the banking crisis that has affected most financial institutions since September 2007, leading to an economic downturn and period of recession. The 'toxicity' of many financial institutions was triggered by excessive lending in a number of countries, including the United States, the United Kingdom, Spain and Ireland, especially on house purchases. This brought about a loss of confidence in the ability of the lending institutions to recoup their assets, and national governments acted to guarantee the private banking sector against a feared 'run on the banks'. These developments have occurred within a context of relatively high personal (and institutional) indebtedness since the 1980s (Ferguson, 2003; Thompson, 2007).

At the same time another shift has been occurring in consumer policy, this time prompted by the much wider acknowledgement of global climate change, especially after the Stern Report was published in 2007 (Stern, 2007). The need to pursue low-carbon solutions to economic growth rapidly altered the policy discourses surrounding consumption, and it has become an article of faith for policy discourse that economic growth is tolerable only if it does not exacerbate existing concentrations of carbon in the atmosphere. For example, in the United Kingdom the Climate Change Bill was introduced in 2008, establishing a very ambitious target for carbon reductions of 80 per cent by 2050. Policy activity has been accompanied by sustained lobbying on the part of NGOs and others, including Rising Tide, the Campaign Against Climate Change, and the series of Climate Camps that have repeatedly mobilised the public to call for urgent action on climate change and a new approach to economic organisation. In effect we have two current policy preoccupations, which are interlinked, but usually considered separately. This book considers how they might be linked and the part they play in moving beyond the conventional discourses of resilience.

Looking for lessons learned from past grand projects has led to several comparisons with experience in Britain during and following the Second World War. War provided a defined and 'external' enemy which facilitated political and class alliances that have not been possible in peacetime (Mackay, 2002). The British experience of austerity provides interesting leads from which to follow the attempts to grapple with the twin challenges of enforced austerity following the fiscal debacle after 2007, and the perceived threats to consumption habits posed by climate change today (Hulme, 2008). In specific ways the challenges of the current fiscal/climate crisis bears comparison with the earlier period. A number of issues may be identified: the role of voluntarism and government enforcement/regulation; the role of propaganda and secrecy, and the extent to which increased citizen participation (rather than that of the market 'consumer') have the potential to alter the relationship between publics and government, and redefine governance. In addition, an examination of the period between 1940 and the 1960s suggests that the credit 'bubble' which burst in 2007 had its origins in the consumer boom of the past half-century, and the climate crisis with which accelerated consumption is linked.

Wartime rationing and austerity represented a very different challenge from the fiscal austerity that is being proposed today. The rise in personal consumption which marked the second half of the twentieth century has served to obscure the experiences of wartime and post-war rationing and scarcity which preceded it (Calder, 1969; Hickman, 1995; Briggs, 2000; Gardiner, 2004; Hennessy, 2006). During the period of austerity, between 1940 and the end of rationing 15 years later, the British people became accustomed to scarcity, to the imposition of administrative edict governing what they could consume and how they could spend (Sissons and French, 1964; Longmate, 1971; Briggs, 1975; Hennessy, 1993). The time line begins with enforced wartime austerity in 1940, when Churchill took over the new Coalition Government, and continues through the impulse of the Beveridge Report (1942) and post-war shortages into the late 1950s and 1960s. The Great Depression of the 1930s had come about as a result of insufficient demand, according to Keynes, and the creation of the Welfare State and the post-war planned economy were attempts to enhance security and increase economic stability through increased demand.

By the 1960s the improvement in household income levels suggested that the model had succeeded. Indeed, the model appeared to have survived the oil crisis of the 1970s, although not without significant modification. In the richer West, the race for economic growth drove privatisation and the hollowing out of the state as decentralisation and globalisation shifted power from public to private domains in the 1970s and beyond. In the poorer, global South, development problems were at first converted into an opportunity for adding value to oil profits through the lending of surplus through international markets or bilaterals. As debt burdens increased economies stagnated or went into decline, with some, especially in Latin America, choosing to default. The global South was then exposed to the harsher side of capitalist development in the 1980s and 1990s, through structural adjustment measures and associated political conditionalities applied by the IMF. These measures were aimed at macro-economic stabilisation but down-played the high social and ecological costs incurred, including the dramatic reduction in public services and associated retrenchment in public sector employment which hit hardest in urban areas as large sectors of the service economy were made ready for privatisation. The result of these trends from the

1960s to the turn of the millennium was a smaller state, economies more open to international investment and large populations of new, geographically mobile and particularly urban poor. The loss of state regulatory power in richer and poor countries alike over this period and encouragement of a globalised financial system can be seen as elements lying at the roots of the current expression of crisis. The co-evolutionary wheel took a significant turn.

The years of austerity following the Second World War laid the basis for modern European consumerism, not simply in seeming to offer choice and individuality, where it had been absent, but in helping to disseminate new social attitudes. In Britain the post-war consensus surrounding the British Welfare State, continued into the 1950s under the soubriquet of 'Butskellism', the hybrid name given by Norman Macrae for the common perspective of R. A. Butler and Hugh Gaitskell, two political leaders who shaped the era (Hennessey, 2006: 211). The Welfare State was constructed on the twin pillars of sound Keynesian economics, and Beveridge's attractive policies for comprehensive social security. It was established to meet need, and succeeded in doing so, but not everyone envisaged it as the launching pad for wholesale advances in personal consumption and credit, such as ensued from the 1960s onward. It was a generation for which 'everything was achievable', and those achievements would not be put fully in jeopardy for another half-century, under the financial crisis heralded by a 'credit crunch', which began in the autumn of 2008.

Thus the austerity associated with the war years was followed by conspicuous over-consumption, since this type of lifestyle was newly available to generations that had grown used to the tight rationing of resources (Soper, 2007). If wartime austerity and its discourses of 'waste not, want not' and 'make do and mend' were ultimately followed by over-consumption, then the use of this kind of approach in the present to generate voluntary sustainable consumption behaviours could conceivably be misplaced, and have even less relevance when the economic recession subsides.

The difficulties in mounting an effective, authoritative and democratic response to the challenges of global climate change today are compounded by the fiscal crisis which has beset most (although not all) the developed economies. General optimism about the economy in the United Kingdom during most of the 1990s, and the escalation in property prices, had served to discourage saving (Bernthal *et al.*, 2005; Braucher, 2006) and increase personal consumption. At the same time the level of indebtedness had increased, even prior to the banking crisis of 2007 to 2008. In much of Western Europe and North America (and especially Spain, the United Kingdom and Ireland) private equity was channelled towards housing and borrowing was easy, individuals were prepared to buy property to rent and re-mortgage their homes with apparent alacrity (Tucker, 1991). More disposable income meant enhanced personal consumption rather than saving, and *sustainable* consumption represented another consumer choice in a buoyant market. It was one way in which the citizen, passenger or neighbour could be relabelled as a 'customer', a discursive practice which had grown since the 1980s, and which drew attention to the ubiquity of market relations (Cross,

1993; Cohen, 2003). For Green and Left critics it also represented a further step towards the privatisation of people's lives and aspirations, and the disarticulation of community and solidarity bonds.

The interest in sustainable consumption was fuelled by the expansion of credit and market opportunities (Bernthal *et al.*, 2005). It consisted largely of widening consumer choice, and making new or ethical products more available on the market, rather than in narrowing choice to fewer, more sustainable products and services. This kind of top-down choice editing has recently been hailed as an important means of delivering sustainable consumption, and would move us away from a reliance on voluntary behaviour change towards greater product selection (Sustainable Consumption Roundtable, 2006).

The rise in disposable income, for most consumers in the richer global North was also driven by increasing female participation in the labour force, facilitating wider market participation for the majority (but not all) of the population (Goodman and Redclift, 1991). This model of rising consumption had also been associated with longer working hours, as Richard Titmuss had argued decades earlier, to explain the apparent rise of the 'Affluent Society' in the late 1950s (Titmuss, 1962) and captured more recently in the concept of 'time poverty' (De Graaf, 2003). In addition, of course, the post-war generation of so-called 'baby-boomers', having paid off their mortgages, had surplus income with which to become further indebted, and property to pass on to their children. This liberal interpretation is not inconsistent with a more Marxian Regulation Theory approach, which seeks to explain the ability of capitalism to stabilise itself in the 1970s and 1980s, but might also help explain the illusion of 'stability' during the long boom of the last decade (Aglietta, 1976; Boyer, 1990; Jessop and Ngai-Ling Sum, 2006). The model of growth at the dawn of the twenty-first century was one of enhanced personal consumption on the basis of negotiated debt.

This model of 'stabilised' debt management and enhanced personal consumption might at first appear at odds with the dominant discourses of sustainable consumption policy, which served as the policy complement of efforts to combat anthropogenic climate change in international negotiations. The emphasis on the individual consumer was in fact quite consistent with the consumer-based policy discourses of the 1990s. The increased purchase of consumer goods and services which carry an 'environmental', 'natural' or 'ethical' imprimatur had been bolted on to a loosely regulated market that prioritised individual choice and profitability over more fundamental shifts in behaviour. The context for most sustainable consumption discourses during the past few years has elements which were consistent with credit expansion and indebtedness, rather than 'self-sufficiency' and deeper Green credentials (OECD, 2002). In fact the sustainable consumption discourses were several, and often mutually contradictory throughout the period in which the idea of Green consumerism as 'sustainable consumption' has become established. They also led, inevitably, to the kinds of path dependency from which modern capitalism cannot easily be extricated, and which have contributed to the wholly inadequate response to the challenge of climate change today.

Global environmental change and the interplay between agency and structure

At the core of the co-revolutionary agenda is the dialectical relation between socio-ecological structures and agency. Socio-ecological factors, such as distribution of entitlements or the class nature of power, structure people's reasons and scope for action. At the same time human agency may shape these structuring factors. The co-creation between structure and agency is key in any attempt to explain social change (Giddens, 1984). Yet, proper acknowledgement of human agency, as a quality of individuals free to act, brings to the table a number of challenges for conceptualising the interaction of individuals, society and nature in a context of global environmental change. First is the question of subjectivity, that human behaviour and environmental attributes are valued differently depending upon viewpoint (Poovey, 1998). This makes any singular objective reality difficult to argue for but also opens analytical leverage on to the problematisation of meanings ascribed to collective challenges as shown in the case of climate change (Hulme, 2008). Second is the need to explore the range of possibilities for agency and its relationship to structure. For instance, agents can be defined as goal-directed, alienated or behavioural; structures can be constraining, enabling or determining; and their interrelation can be unidirectional, bi-directional or co-evolving. The problem is not to find the 'right' definition but the fact that, so far, this ontological discussion hardly ever takes place in global environmental change literature.

A promising entry point for the serious consideration of agency within global environmental change literature is the development of theories on self-organising systems, and, in particular, Prigogine's theory of dissipative structures (Nicolis and Prigogine, 1977). This theory shows that matter and life are capable, within the limits of deterministic physical laws, of producing new patterns of organisation and 'doing things' – for example in the emergence of tornadoes or hurricanes – causing other subjects and physical or social patterns and behaviours to change and adapt. This is part of the mechanism of co-evolution – of the interaction of the physical and social in ways that make it increasingly difficult to define a nature that is out there acting independently of social construction as well as a social world that is independent of nature (Castree, 2005). As natural elements are recognised to exhibit many aspects previously attributed only to human actors one moves closer to a conceptualisation of society-nature as an integrated socio-ecological system where elements can self-organise as they co-evolve.

As discussed in the preceding section, modern capitalist societies tend to conflate human agency with consumer roles. This leads to portraying global environmental change response as a matter of affecting consumer choices. More generally, consumerism has come to provide a meta-framework through which we understand our relationship with the material world of nature. This reductive frame plays down understandings of human agency that emphasise co-dependence with ecological and biological processes, spirituality, or altruistic commitment with others (Manuel-Navarrete *et al.*, 2004). In this context one may argue that global environmental change research is critical in being able to challenge assumptions about human agency based in the contemporary consumerist *Zeitgeist* and the rising hegemony of market logics and Western liberalism in its several guises. From this critical ontological position, participation in markets is to be seen as just one of the ways in which collective action is shaped, and likely not the most effective when it comes to the transformation of socio-political structures.

From the point of view of the combined contemporary crisis of financial and climate crises, the analysis of political agency has still to deal, first, with the issue of the conditions under which the stern resilience of entrenched structures may be effectively overcome. Multiple voices have recently been heard in demand of structural changes, especially during the 2009 climate change negotiations in Copenhagen, or even more so during the worst of the global financial crisis. However, there has been remarkably little change, even in the sense of shy reforms, in the structures that led us, and may still be leading us, into these crises in the first place. In our view, this absence of actual change poses several pressing research questions: Is structural change prevented by the constraining that these very structures exert over individual agents, who are thus unable to draw on their own creative capacity for change? Which are the main bottlenecks? What do we know about, for instance, agents' ability to translate introspective motivations or reflexive power into actual transformations of entrenched structures?

Second, there is room for the analysis of changes in human agency itself or at least in dominant forms of agency as they are promoted from particular social structures. Arguably, this analysis is crucial for exploring the possibility of gearing global structures towards ecological sustainability and global equity. Socio-ecological agency has been proposed as a concept to explore changes in human agency in the context of global environmental change (Manuel-Navarrete and Buzinde, 2010). It is argued that the confluence of multiple global crises may be heading us into framing human agency in terms of global stewardship. Socio-ecological agency is to be exercised by individuals who acknowledge co-dependency and co-evolution between society, the environment and their inner worlds. Depicting agency as global stewardship calls for the individual and collective self-imposition of material constrains for the sake of the planet's integrity and wealth redistribution. However, these voluntary self-limitations cannot be just embraced by some individuals or specific societies. If they are to have any significant effect, they must be exercised by humanity as a whole. Yet, framing this challenge in terms of coercion and a governance problem runs the risk of missing that at stake here is the very logics of current governance structures which cling tightly to the illusions of never-ending material growth and the trickling down of wealth. This book discusses the limits of ecological and economic reform within capitalism as well as the possibilities and limits of transforming both socio-political structures and human agency in the context of cumulative socio-ecological crises.

Conclusion

This chapter began the task of defining the underlying causes of the current combined socio-economic, political and ecological crisis. Co-evolution reveals the roots of contemporary causes in historical process – and the need to break such cycles of creative destruction. The two dominant features of the current capitalist model – economic globalisation and the carbon economy – have persisted because they have promised to offer much: a reduction in inter-state conflict, especially in the global economic core and increasing levels of welfare worldwide. But any achievements are accruing at increasingly high costs for human welfare – indirectly through global environmental change – and are felt more directly by the many millions who remain marginalised from the gains of development currently structured. A second feature of co-evolution is the self-reinforcing lock-in that these two aspects of the global economy have instilled, making any purposeful change in direction difficult. Globalisation is defined by a shift in power from nation states to footloose capital with concomitant reduction in regulation and accountability such that states are now less able to act to redress imbalance (as we have seen in the stalemate of WTO and UNFCCC negotiations). This leaves the market and civil society as the prime 'actor-systems' available to make adjustments for sustainability. Social and environmental externalities have long been identified as weaknesses that limit the appropriateness of market solutions. Simultaneously, the carbon economy has transformed everyday life touching upon our very senses of identity, which have become increasingly associated with material consumption, and perhaps more profoundly the very systems that provide food (through chemical fertilisers and international transport), heat and increasingly (through desalination plants) even water for the rapidly growing and urbanising global population.

Given the strength of lock-in which few individuals are in a position to voluntarily transcend (though many more are locked out through poverty), where might we begin to look for points of opportunity and mechanisms for change? Certainly it is important to reflect on past framings of socio-ecological tension. Michael Redclift offers just such an assessment of sustainable development discourses seen through the lens of the current crisis. Contemporary policy frames also need close scrutiny, and the chapters by Katrina Brown and Mark Pelling test the contribution of resilience as an overarching discourse, the former exploring its use and meaning in international development, and the latter suggesting a more nuanced interpretation that can better emphasis transitional and transformational alongside resilient futures. Andy Gouldston and Rory Sullivan argue that ecological modernisation offers some prospect but is limited by its positioning within dominant features of the global carbon economy. However, still the urgency of responding to climate change and the apparent inability of political and spontaneous civil action requires close attention to market-driven, technological solutions – even if these are only partial. Elsewhere, more profound alternatives are so far limited in scope; two are diagnosed here: Diana Mitlin explores capacity and lessons for political and environmental change that can be learned from the local successes of urban peoples' movements in Asian cities, and Peter North and Molly Scott Cato examine the social context and drivers for the Transition Towns movement in the UK that seek more local control over energy and economy.

The search for points of engagement leads Marcus Carson to examine the mechanisms of political decision-making that have contributed to the lock-in of unsustainable development in rich countries. Both John Barry and David Manuel-Navarrete then develop this argument by tackling the notion of economic growth, Barry at a macro-economic scale and Manuel-Navarrete through the lens of individual life histories. The final analytical chapters of this collection return to an examination of the mechanisms that frame decision-making, but move from an analysis of framing of the present to the framing of possible future politics of climate change. Mattias Hjerpe and Björn-Ola Linnér offer an analysis of the potential for utopian thinking and its place in scientific and policy responses to climate change. Ian Bailey and Hugh Compston make clear the relational and systemic challenges that will be part of any new politics.

These chapters cover a wide ground and one that would be difficult for an individual scholar to master. Together they offer diagnosis and also point to opportunities for making fundamental shifts in culture, economy, the structure of politics and everyday life that together point towards scope for David Harvey (2010)'s call for a co-revolutionary agenda for not only survival but human welfare and flourishing.

References

Aglietta, M. (1976) *A Theory of Capitalist Regulation: The United States Experience*, London: Verso.

Beck, U. (2008) *World at Risk*, Cambridge: Polity Press.

Beck, U. (1992) *Risk Society*, London: Sage.

Beck, U., Giddens, A. and Lash, S. (1994) *Reflexive Modernization: Politics, Tradition and Aesthetics in the Modern Social Order*, Cambridge: Polity Press.

Bernthal, M., Crockett and Rose, R. (2005) Credit cards as lifestyle facilitators. *Journal of Consumer Research* 32 (1): 130–145.

Braucher, J. (2006) Theories of over indebtedness: interaction structure and culture. *Theoretical Inquiries in Law* 7(2).

Bowers, S.R. (1993) Soviet and post-Soviet environmental problems. *Journal of Social, Political and Economic Studies* 18 (2): 131–158.

Boyer, R. (1990) *The Regulation School: A Critical Introduction*, New York: Columbia University Press.

Briggs, A. (2000) *Go To It! Working For Victory on the Home Front 1939–1945*, London: Mitchell Beazley/Imperial War Museum.

Briggs, S. (1975) *Keep Smiling Through: The Home Front 1939–1945*, London: George Weidenfeld and Nicolson.

Calder, A. (1969) *The People's War: Britain 1939–1945*, London: Pimlico.

Castree, N. (2005) *Nature*, London: Routledge.

Chatterjee, D.K. (2007) *Democracy in a Global World: Human Rights and Political Participation in the 21st Century*, New York: Rowman & Littlefield.

16   *M. Pelling* et al.

Cohen, L. (2003) *A Consumer's Republic: The Politics of Mass Consumption in Postwar America*, New York: Vintage.

Cross, G. (1993) *Time and Money: The Making of Consumer Culture*, New York: Routledge.

De Graaf, J. (2003) *Take Back Your Time: Fighting Overwork and Time Poverty in America*, San Francisco, CA: Berrett-Koehler.

Dicken, P. (2007) *Global Shift: Mapping the Changing Contours of the World Economy.* (5th edn) London: Guilford Press.

Eckersley, R. (2007) Ambushed: the Kyoto Protocol, the Bush administration's climate policy and the erosion of legitimacy. *International Politics* 44: 306–324.

Ferguson, N. (2003) The nation; true cost of hegemony: huge debt. *New York Times*, 20 April.

Fromm, E. (1961) *Marx's Concept of Man*, New York: Frederick Ungar.

Gardiner, J. (2004) *Wartime: Britain 1939–1945*, London: Headline Book Publishing.

Giddens, A. (1984) *The Construction of Society*, Cambridge: Polity Press.

Giddens, A. (2009) *The Politics of Climate Change*, Cambridge: Polity Press.

Goodman, D.E. and Redclift, M.R. (1991) *Refashioning Nature, Food, Ecology, Culture*, London: Routledge.

Gorz, A. (1980) *Ecology as politics*, London: Pluto.

Handmer, J.W. and Dovers, S.R. (1996) A typology of resilience: rethinking institutions for sustainable development. *Organization and Environment* 9 (4): 482–511.

Harvey, D. (2010) *The Enigma of Capital*, London: Profile.

Hennessy, P. (1993) *Never Again: Britain 1945–1951*, London: Vintage.

Hennessy, P. (2006) *Having It So Good: Britain in the Fifties*, London: Allen Lane.

Hewitt, K. (ed.) (1983) *Interpretations of Calamity*, London: Allen & Unwin.

Hickman, T. (1995) *What Did You Do in the War, Auntie? The BBC at War 1939–1945*, London: British Broadcasting Association.

Hulme, M. (2008) The conquering of climate: discourses of fear and their dissolution. *The Geographical Journal* 174 (1): 5–16.

Hylton, S. (2001) *Their Darkest Hour: the hidden history of the Home Front 1939–1945*, Stroud: Sutton Publishing.

Jessop, B.J. and Ngai-Ling Sum (2006) *Beyond the Regulation Approach*, Chichester: Edward Elgar.

Klein, N. (2007) *The Shock Doctrine: The Rise of Disaster Capitalism*, London: Penguin.

Longmate, N. (1971) *How We Lived Then: A History of Everyday Life During the Second World War*, London: Hutchinson.

Mackay, R. (2002) *Half the Battle: Civilian Morale During the Second World War*, Manchester: Manchester University Press.

Manuel-Navarrete, D. (2010) Power, realism, and the humanist ideal of emancipation in a climate of change. *Interdisciplinary Reviews: Climate Change* 1 (6): 781–785.

Manuel-Navarrete, D. and Buzinde, C.N. (2010) Socio-ecological agency: from 'human exceptionalism' to coping with 'exceptional' global environmental change. In Redclift, M.R. and Woodgate, G. (eds) *The International Handbook of Environmental Sociology* (second edn) Cheltenham: Edward Elgar, pp. 306–337.

Manuel-Navarrete, D., Kay, J.J. and Dolderman, D. (2004) Ecological integrity discourses: linking ecology with cultural transformation. *Human Ecology Review*, 11 (3): 215–229.

Meyer, J.W. and Jepperson, R.L. (2000) The 'actors' of modern society: the cultural construction of social agency. *Sociological Theory* 18 (1): 100–120.

Millennium Ecosystem Assessment (MEA) (2005) *Ecosystems and Human Well-being: Synthesis Report*, Washington, DC: Island Press.

Nicolis, G. and Prigogine, I. (1977) *Self-organization in Nonequilibrium Systems: From Dissipative Structures to Order Through Fluctuations*, New York: Wiley.

Norgard, R.B. (2004) *Development Betrayed: The End of Progress and a Co-evolutionary Revisioning of the Future*, London: Routledge.

Norgaard, R.B. (1995) Beyond materialism: a coevolutionary reinterpretation of the environmental crisis. *Review of Social Economy* 53 (4): 475–492.

Organisation for Economic Cooperation and Development (OECD) (2002) *Towards Sustainable Household Consumption? Trends and Policies in OECD Countries*, Paris: OECD.

Pelling, M. (2003) Paradigms of risk. In Pelling, M. (ed.) *Natural Disasters and Development in a Globalizing World*, London: Routledge.

Pelling, M. (2010) *Adapting to Climate Change: From Resilience to Transformation*, London: Routledge.

Poovey, M. (1998) *History of the Modern Fact: Problems of Knowledge in the Sciences of Wealth and Society*, Chicago, IL: University of Chicago Press.

Schneider, S.H. and Londer, R. (1984) *Coevolution of Climate and Life*, San Francisco, CA: Sierra Club Books.

Sewell Jr., W.H. (1992) A theory of structure: duality, agency, and transformation. *American Journal of Sociology* 98 (1): 1–29.

Sharer, R.J. (2006) *The Ancient Maya* [6th edn], Stanford, CA: Stanford University Books.

Sissons, M. and French, P. (1964) *The Age of Austerity*, London: Penguin Books.

Smith, C. (2005) *Karl Marx and the Future of the Human*, Lanham, MD: Lexington Books.

Smith, N. (2010) *Uneven Development: Nature, Capital and the Production of Space*, London: Verso.

Soper, K. (2007) Re-thinking the 'good life': the citizenship dimension of consumer disaffection with consumerism. *Journal of Consumer Culture* 7 (2): 205–229.

Sotarauta, M. and Srinivas, S. (2006) Co-evolutionary policy processes: understanding innovative economies and future resilience. *Futures* 38 (3): 312–336.

Stern, N. (2007) *The Economics of Climate Change: The Stern Review*, New York: Cambridge University Press.

Sustainable Consumption Roundtable (2006) *I Will if You Will: Towards Sustainable Consumption.*

Titmuss, R. (1962) *Income Distribution and Social Change*, London: Allen & Unwin.

Tucker, D. (1991) *The Decline of Thrift in America: Our Cultural Shift from Saving to Spending*, New York: Praeger.

Part I
Problem framing

2 Living with a new crisis

Climate change and transitions out of carbon dependency

Michael Redclift

Introduction

This chapter examines the background to the current financial crisis and the problems surrounding policies to combat climate change through transitions out of dependence on carbon. After providing a critique of the current situation the chapter suggests that there are fundamental flaws in the way in which policy has addressed both agency and structure in relation to climate change. It argues for a need to draw away from the path dependence that helped to define mainstream policy initiatives focused on individual consumer behaviour, and argues for a stronger recognition of structural inequalities, internationally and nationally, as the cornerstone of an alternative Green political stance. It suggests that we need to look again at 'post-carbon' futures from within the compass of globalisation and its consequences. Globalisation has accelerated the risks of climate change through ushering in new capitalist economies, notably in China and India. It has also appeared to offer new certainties – global economic interdependence and the harnessing of nature for the purposes of economic development – that provide part of the 'problem' of climate change itself. These new geopolitical and cultural realities have made it difficult for the critics of globalisation to move much beyond critique. If we have now arrived at a 'tipping point' on climate change (IPCC 2007), we need to address the problem of decarbonisation through an approach that goes well beyond market 'mechanisms', and requires social and political mobilisation (Redclift, 2010; Sachs 2010). Figure 2.1 illustrates the synergistic relationship between globalised environmental and economic systems, and with policy and practice domains where behavioural shifts may challenge dominant development visions/policy.

Historically, the last few decades of the twentieth century witnessed a series of crises over the economy and the environment. The apparent incompatibilities between economic growth and environmental sustainability came to mean two things:

1   On one account the limits of resource capacity were in danger of being exceeded. Resource shortages were a constraint on further economic growth and development. We should conserve resources to facilitate growth. This was essentially the 'Limits to Growth' position in the early 1970s (Meadows *et al.* 1972).

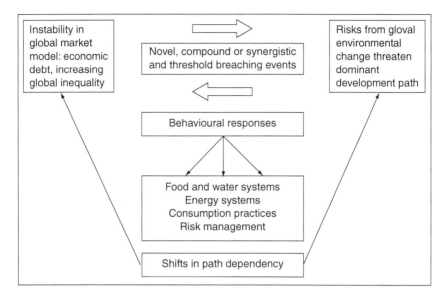

Figure 2.1 From crisis to opportunity.

2   Existing levels of economic growth represented a threat to the environment and resources – a vicious circle had been created in which economic activity undermined the biospheric resources on which we rely. (This is the 'weak' Green position, usually referred to as 'sustainable development', which evolved largely since the 1980s. A 'stronger' view of sustainable development would imply much lower levels of substitution between human-made and natural capital, and corresponding discount rates) (Pearce 1991).

The first position lost support partly because it was a product of high energy prices (the oil hikes of the 1970s). As hydrocarbons became relatively cheaper, and the effects of the Green Revolution in expanding food staples to meet population growth began to be acknowledged, it was no longer clear that the Malthusian position held – that population exceeded the resources necessary to feed this growth. In addition, the drive for economic development in the South (circa the 'Brandt Report') was overtaken by events (Brandt 1980). At first, it was placed in jeopardy by the debt crises of the 1980s, the structural adjustment programmes, and 'post-recovery', the deregulation of markets, the retreat of the state and, eventually, higher levels of economic growth in much of the newly developing world, especially the populous economies of Asia.

The genius of the second position ('sustainable development') was that almost everybody could sign up to it. There were very few dissenting voices (Redclift 1987; Norgaard 1988; Adams 1990) The mechanisms which were unleashed via deregulation and the neoliberal ascendancy ('The Washington

Consensus' – a 'consensus', incidentally, in which most people had not been consulted) became the favoured instruments of policy in seeking to achieve 'sustainable development' (Stiglitz and Serra 2008). These took two forms.

First, attempts were made to internalise environmental externalities in products and services – 'Ecological Modernisation'. This was viewed as a competitive strategy by the European Union in particular, giving Europe a competitive advantage over the United States and any newly developing rivals. Basically, you count the embodied carbon in products, seek to reduce energy and material throughput, and make a 'win/win' gain, by reducing energy costs (hydrocarbons prices were rising) and reducing environmental damage. Trade arrangements would also take account of 'embodied carbon'. The more interventionist policies of the European Union facilitated this in the 1990s.

Second, the development of carbon markets, both within industries and, more importantly, between countries. These new markets represented a challenge for entrepreneurship, new market opportunities, and required very little government action. Carbon markets were thus popular among devotees of free-market economics and environmentalism, unlike other interventions such as carbon taxes (Simms 2005) (although few paused to consider what might happen when markets fall and the price of carbon dropped significantly). Both of these developments are examined later in the chapter.

The conversion of governments to a more or less uncritical view of markets was even more evident in the international efforts to 'protect' biodiversity. The biodiversity regime was expressed in the Convention on Biological Diversity (1992) and the Cartagena Protocol on Biosafety (2000). This demonstrated a shift from a focus on the loss of *species* diversity, and thus the loss of complex ecosystems to a focus on the preservation of *genetic* diversity, where the principal gains were in the pharmaceutical industries and agriculture (Paterson 2008). Again, the almost imperceptible shift was from *nature conservation* to *nature as commodity*. The main opposition to the latter was from groups – mainly NGOs – which argued that marginalised people had *rights* in nature which governments and the pharmaceutical industry ignored. However, the industry lobby won much of the ideological struggle, insisting that *ex situ* conservation in gene banks should be treated as equivalent to *in situ* conservation in ecosystems.

From a Left perspective the conjunction of newly 'liberated' markets and environmental concern was a necessary contradiction of capitalism seeking a resolution, and could with hindsight be seen as a 'managed senescence', if we continue with the biological metaphors of 'development' (Smith 2007; Woodgate 2010; Bellamy-Foster 2010). A more mainstream view, however, would be that they addressed system failures, and could even lead to a rejuvenated, if scarcely recognisable type of materials 'light' capitalism (Lovins *et al.* 2000).

A novel crisis: financial markets and the environment

The hopes that markets and technology, together with more informed personal choices, would solve the environmental problems associated with accelerated economic growth and the enormous rise in global consumption were about to be challenged by a number of events:

1   The financial crisis which began in 2007/2008 was a 'crisis' fed by the institutionalised but ultimately personal greed of many bankers and financial managers, and fuelled by the virtually unregulated production of credit – not because interest rates were low, but because the price attached to housing equity (especially in the United States, the United Kingdom, Spain and Ireland) was unrealistically high. The rise in 'sub-prime' lending and borrowing, took place under systems of ineffective governance which emphasised everybody's right to property regardless of collateral and debt levels. Politically it was 'sold' to consumers as everybody's right to credit rather than their right to debt. The financial crisis revealed that the model was completely unsustainable.

2   The policy response paid lip-service to the rapidly disappearing Green policy agenda, but did not support this rhetoric with effective interventions (cf. the almost derisory role of new Green investment in attempts to address the financial crisis).

3   There was also now considerable evidence of the effects of the financial downturn on migration, as well as poverty, and on a continental scale, notably in China, which supported the United States' debt through buying into its financial packages, and supported raised consumption in the West generally, by lowering the costs of manufactured goods there. It was estimated that, despite the global economic recession, internal migration between the interior and the coast of China continued to accelerate, while China maintained an impressive annual growth rate of almost 10 per cent. The large, newly developing economies, especially in Asia, were increasingly looked to for the stimulus which would lead the economies of Europe and North America out of recession.

4   Another process that has gathered speed is that of the transnational sourcing of food, minerals and other resources. The internationalisation of capital movements and the need to secure resources led to increased transnational acquisition of land and minerals on the part of China and some of the Gulf states, principally in Africa. Rather than depend exclusively upon trade relations to meet their domestic resource deficiencies – since trade contracts during an economic recession – the advantages of acquisition of land, water sources and food (via 'virtual water') became evident, especially for their geopolitical reach (Allan 2003). Land displacement for crops like soya had already changed international food/land imbalances, but the leasing and ownership of land in South/South transactions was new.

Each of these processes appeared to have undermined the international economic stability that had been taken for granted, with a few wobbles, since the 1980s. They also suggested connections between the financial and environmental crises that had been only dimly perceived in the past. Other questions remained relevant: could one generalise a 'successful' model of economic development to the global scale without causing irreparable harm to the environment? And, in pursuing international policies to limit the damage of climate change (Copenhagen 2009), were individual country economies being placed in jeopardy?

The financial crisis, climate change and consumption

The changes in the way in which materials, food and energy are sourced globally have usually been discussed without much reference to sustainable consumption. Today the expansion of credit in much of the developed world, and the associated levels of personal and corporate debt, are necessarily linked with the banking crisis that has affected most financial institutions since September 2008, leading to an economic downturn and period of recession. An understanding of the 'limits' imposed by shifts in demand needs to be complemented by an analysis of the rising levels of personal consumption and debt, not only in the developed world but in many middle-income and fast-growing developing economies (Durning 1992; Redclift 1996; Princen *et al.* 2002). As we shall see, these issues are closely linked to the global problems surrounding climate change, and are manifested at different spatial scales.

The 'toxicity' of many financial institutions was triggered by excessive lending, and low interest rates prompted by the need to finance the war in Iraq (*The Economist* 23 January 2009). Financial institutions linked to housing equity in a number of countries helped to create an unrealistic credit profile, especially in the United States, the United Kingdom, Spain and Ireland. The rise of the credit economy, as well as the popularisation of new instruments like home equity loans with which to draw on expected capital gains, contributed to dramatic changes in commercial credit markets. There are clear indications from the literature over a decade ago that savings behaviour shifted most sharply in countries with more liberal access to personal credit, notably the United States and the United Kingdom (Calder 1999; Parker 1999; Manning 2001; Guidolin and La Jeunesse 2007). As we have seen, this brought about a loss of confidence in the ability of the lending institutions to recoup their assets, and national governments acted to guarantee the private banking sector against a feared 'run on the banks'. These developments occurred within a context of relatively high personal (and institutional) indebtedness since financial deregulation was initiated in the 1980s.

At the same time another shift has been occurring in consumer policy, this time prompted by the much wider acknowledgement of global climate change, especially after the Stern Report was published in 2007 (Stern 2007). The need to pursue 'low-carbon' solutions to economic growth rapidly altered the policy discourses surrounding consumption and the environment, and it has become an article of faith for public policy that economic growth is tolerable only if it does not exacerbate existing concentrations of carbon in the atmosphere. In 2008 the United Kingdom's Climate Change Bill was introduced, establishing a very ambitious target for carbon reductions of 80 per cent by 2050. This policy activity has been accompanied by sustained lobbying on the part of NGOs and others, including Rising Tide, Friends of the Earth, the Campaign Against Climate Change, and the series of Climate Camps that have repeatedly mobilised sections of the public.

This new policy perspective is seen clearly in the document which, more than any other, represents the high-watermark of free-market environmentalism. The Stern Review (2007) noted that:

> The transition to a low-carbon economy will bring challenges for competitiveness but also opportunities for growth.... Reducing the expected adverse impacts of climate change is therefore both highly desirable and feasible.
>
> (Stern 2007: 32–33)

This quotation illustrates the way in which what had previously been viewed as a 'threat' could quickly become an 'opportunity', although the quotation fails to say for whom the opportunities exist. Unsurprisingly, the immediate responses to Stern (and the IPCC Fourth Assessment of 2007) were optimistic in tone. One commentator on business and the environment wrote:

> People would pay a little more for carbon-intensive goods, but our economies could continue to grow strongly.... The shift to a low-carbon economy will also bring huge opportunities.... Climate change is the greatest market failure the world has seen.
>
> (Welford 2006: 261)

The characterisation of climate change as a 'market failure' immediately offered economists, businesses and government a lifeline. Rather than necessitating expensive and comprehensive restructuring in new systems of provision, or even reduced volumes of production and consumption, Stern's neoclassical view was that sustainability could be delivered through *increased* consumption of particular kinds of products simultaneously. Feeding the economy has come to typify the mainstream environment and consumption discourse.

There have been no significant shifts in this respect since the UK Coalition government came into being in the summer of 2010, at least in the emphasis that continues to be placed on utilising market instruments and voluntarism to boost Green objectives, whether through subsidising the private acquisition of solar panels, the commitment to more investment in nuclear power or the drive to recycle more household waste. One significant shift has been away from what some saw as New Labour's 'nanny state' approach (that is, the encouragement of individual shifts in behaviour) towards less government 'interference'. The unwillingness of the new government to 'interfere', it must be said, has also been prompted by the size of the UK fiscal deficit. Prime Minister David Cameron's boosting of the 'Big Society', third sector and private sector organisations, at the expense of the state, plays uncomfortably with the long-term strategic objectives of Green policy to make massive reductions in carbon emissions, since it remains unclear how and when the necessary changes will take place.

These developments in the economy and in public policy raise some awkward questions for our understanding of the policy discourses which have characterised the field. There is still considerable confusion over the most effective way of reducing consumption and the accompanying carbon emissions, and several of the assumptions about consumer behaviour – such as the role of an 'information deficit' surrounding the environmental costs of products and services – are, at best, questionable (Redclift and Hinton 2008). Remarkably, assumptions about personal behaviour being triggered by available information are also largely untested. While policy-makers and pundits alike tend to measure progress towards sustainable consumption in terms of the numbers of purchases of particular 'green' or 'ethical' commodities, where success is framed in terms of market share, an alternative discourse suggests that sustainable consumption rests on other facets of behaviour such as frugality, thrift and a kind of voluntary austerity (Soper *et al.* 2008). If this is indeed the case, then a focus on lowering economic growth may still be preferable to pursuing 'sustainable growth' strategies.

As the quote from the Stern Report above suggests, climate change is now regarded as a 'given', and markets are now considered more relevant to policy solutions than ever before. The reduced dependency on hydrocarbons is widely regarded as deliverable through changing consumer policy. The language of 'Green consumerism' can reduce the politics of climate change to the size of a Green consumer product. The experience of an economic recession in the developed world has perhaps served to intensify this process, creating policy tensions where once there were only 'policy opportunities'.

This chapter began by arguing that the 'contradictions' of thinking about sustainability and development have merged into distinct policy discourses, around the idea of 'natural limits', resource capacity and (un)sustainable consumption. Each of these discourses can be usefully informed by recent work in the social sciences which explores the changing role of science policy. A realist, science-driven policy agenda has been paralleled by a science-sceptical postmodern academic discourse (Yearley 1996; Demeritt 1998). Neither position represents a threat to the other – since they inhabit quite different epistemological terrain, and address different audiences. In the process, however, we have seen an enlarged academic debate, and one that closely examines the way in which environmental language is deployed, while at the same time recognising that public policy discourses themselves carry weight. However, issues around the social authority of science, and the way it is employed politically, are linked to structural shifts both in the 'formal' economy and in the 'informal' and more recent 'virtual' economies. These connections are examined below.

The policy debate has proceeded through assumptions about 'choice' and 'alternatives' that have been largely devoid of any critical, structural analysis, and frequently narrow the sociological field of opportunity, by assuming that people act primarily as consumers rather than as citizens (Redclift 2010). There is clearly room for more rigorous analysis of what is a very broad social terrain beginning with the assumptions of the model implicit in most environmental and consumer policy.

Assumptions of the policy model

The neoliberal model which characterised the 1980s and 1990s was viewed by many as a liberating model. It removed 'government' as the engine of economic momentum and opened up activities to the market, or introduced 'shadow' markets which encouraged individuals to behave as if markets operated, in the process not merely shifting economic activities to the private sector but implementing a new logic for the public sector (a sector which continued to grow in most developed countries). The new policies also deregulated financial flows, facilitating the movement of capital (particularly finance capital), and lessened the burden on capital through reducing penalties on growth such as corporate taxes. The model also removed many of the politically negotiated rights that organised labour had gained in the developed world, and reconfigured the frontiers of the 'welfare state'. Among the existing capitalist economies, only those of the European Union sought to combine this market-based model with measures in favour of labour, consumer and environmental protection, producing a hybridisation of neoliberal thinking and traditional welfare support.

Rethinking the role of the state and the consumer in economic growth held importance for the environment, too. The new policy emphasis, especially within the European Union, was on moving from the management of capitalist growth in line with sustainability, towards enabling private actors to pursue their interests while *simultaneously* promoting sustainability. Policy sought increasingly to structure incentives for actors, believing that the agency of the individual, if it existed at all, consisted of a kind of consumer agency, rather than the complex roles that constitute citizenship. This wider view of the multifarious roles performed by the citizen had been pioneered by social democratic (and some Christian Democrat) governments. However, the new model envisaged the individual as their 'consumer self', and this applied as much to the way in which environmental externalities were treated as to the loosening up of credit, and (in the case of some economies) the model of economic growth which was predicated on expanding credit (and with it indebtedness) using the equity of the 'homeowner' as the principal collateral.

These changes came at a cost, of course. The movement of neoclassical economics into more mainstream environmental policy left several concerns at the margin of policy and politics. The challenges of reducing material throughput, and reducing carbon emissions, converted environmental policy into a technical question, while the agency of social movements and their pursuit of alternative social and cultural objectives was effectively sidelined. Unlike the position in the first half of the twentieth century, for the discursive politics of the decades after 1980 the term 'utopia' was treated pejoratively, as irrelevant and out of phase with the realities of the 'enabling market'. The apparent need to reassure publics that the impending environmental dystopias were not inevitable seems to have led policy-makers to emphasise individual contributions ('every little helps') above collective political action. Unlike other policy domains, notably education, the environment was not considered 'aspirational' by many leading politicians.

The underlying assumptions of the dominant model transposed the supposed 'barriers' to market freedom and choice in the formal economy, to the new terrain of environmental and sustainability policy. Policy interventions assumed that similar barriers, this time social rather than economic, existed to people acting more sustainably in everyday life (Redclift and Hinton 2008). These social barriers were constituted by habit, poor education, and a lack of information and state bureaucracy, and could be rectified by policy. The solution was to introduce more choice of products and services, new 'Greener' technologies, and market opportunities which could maximise utility while placing more responsibility on the individual. This solution rendered the environment solely in terms of products and services, rather than social process or structure.

At the same time science was viewed as part of the solution, rather than the 'problem' confronting societies threatened by climate change. The decisions were only obliquely political, and technical solutions held the promise of removing politics from environmental policy entirely. As demonstrated in the Stern Report, we were embarking on what has been termed a 'post-political' future (Swyngedouw 2009): one in which consensus science came to exercise normative authority, and political judgements about the way in which resources and rights to them were distributed could be left to (supposed) independent rational discussion.

One illustration of this new approach to the primacy of the individual consumer is the division of the population favoured by DEFRA (2008) in the United Kingdom. According to this model of British society, and drawing on surveys conducted among British consumers, there are seven kinds of consumers exercising their choices in the United Kingdom today (see Box 2.1).

This market research approach to modelling consumer behaviour regards consumer attitudes as proxy for social and economic structures. It matters little whether a consumer is a poor single parent living in a high-rise housing complex or a wealthy household living in a rural area, using two cars to do the shopping and ferry the children to school. What matters is that the attitudes displayed influence the household's level and type of market engagement. The task then is to tailor policy to different consumer profiles. It is largely irrelevant whether society changes as long as behaviour does.

Capitalism 'lite'

At a more 'macro' level the development of carbon markets, both within industries and, more importantly, between countries, represents a mature version of the 'market solution' model. On the one hand, the development of carbon markets was welcomed by many sectors of industry; indeed, they were heralded as a 'challenge for entrepreneurship', providing new 'market opportunities' (Lovins *et al.* 2000). At the same time, as we have seen, they required very little government action, and were consistent with the largely deregulatory model being widely pursued.

Box 2.1 DEFRA's division of the UK population by consumer 'segments'

1 **Positive Greens:** 'I think it's important that I do as much as I can to limit my impact on the environment' (18% of survey participants).

2 **Waste watchers:** 'Waste not, want not, that's important, you should live life thinking about what you are doing and using' (12%).

3 **Concerned consumers:** 'I think I do more than a lot of people. Still, going away is important, I'd find that hard to give up.... Well I wouldn't, so carbon offsetting would make me feel better' (14%).

4 **Sideline supporters:** 'I think climate change is a big problem for us. I know I don't think much about how much water or electricity I use, and I forget to turn things off... I'd like to do a bit more' (14%).

5 **Cautious participants:** 'I do a couple of things to help the environment. I'd really like to do more, well as long as I saw others were' (14%).

6 **Stalled starters:** 'I don't know much about climate change. I can't afford a car so I use public transport... I'd like a car though' (10%).

7 **Honestly disengaged:** 'Maybe there'll be an environmental disaster, maybe not. Makes no difference to me, I'm just living life the way I want to' (18%).

Source: Defra (2008)

Carbon markets were thus popular among devotees of free-market economics and those who recognised the urgency of environmental action, but who bemoaned the shifts in behaviour this might imply. As one 'progressive' think-tank in the UK put it, 'they (provide) the political opportunity to highlight, secure and celebrate wealth creation. The benefits from the low-carbon transition are waiting to be grasped' (Policy Network 2008: 23). Notwithstanding the endorsement of carbon markets by large sections of political opinion, they also raised other questions which were anathema to more radical Green opinion, heightening the possibility, following Oscar Wilde's famous dictum, of 'knowing the price of everything and the value of nothing'.

The existence of carbon markets contributed to the new middle-ground consensus that has come to characterise business-friendly environmental policy during the first decade of the twenty-first century. Organisations like the Carbon Trust advertised heavily in publications such as *The Economist*, where individual entrepreneurs were singled out for compliments and communicated their endorsement of carbon trading. 'What was I thinking when I cut our carbon and joined the standard?' asks Chris Pilling of HSBC. The answer is a conclusive 'win/win' piece of advocacy: 'I saved money, gained a competitive edge, improved efficiency and shared the tangible benefits of accreditation.'

The clear benefits of encouraging industry to enter the new carbon markets represented only one part of the equation however. The downsides of carbon trading were perhaps less 'tangible' but equally compelling. Once the financial recession became apparent the benefits of carbon *markets* began to recede.[1] By late 2009 the 'cap-and-trade' model was beginning to lose ground in precisely those economic systems which had earlier favoured it. Under the legislation being prepared in the United States, under President Obama, electricity utilities looked likely to use 'cap and trade' but transport emissions were more likely to be taxed and industrial emissions regulated. The 'new tools' of the market was less in evidence in 2010 than ten years earlier. By the same token the appeal of the 'old' policy instruments of taxation and regulation was more apparent during 'bad times' when governments, especially in the United States and Europe, needed to raise income, particularly for much-needed new investments in energy (including renewables). As *The Economist* put it: 'climate action may come to lean more heavily on the command-and-control techniques than the market-based approaches they were intended to replace' (*The Economist*, 20 March 2010).

What is the significance of carbon markets for individual consumers, whose attention has increasingly been drawn towards ways of reducing carbon 'footprints': the mechanism favoured by many mainstream commentators? Carbon footprints appear to provide a ready-made and measurable way of enabling individuals to make choices about travel in particular, leading some of them to 'offset' some choices against others and improve their sustainability 'profile'. This has led some commentators to advocate individual carbon budgets as the logical consequence of carbon measurement.

However, there are a number of problems associated with carbon footprinting that are not always discussed. First, although it is a technique which allows comparisons between indicators, carbon footprints cannot be converted into monetary or social values, so are of only limited use to policy (OECD 2004; Schmidt 2009). In addition, measuring an individual's carbon footprint does not help us to understand what an acceptable rate of carbon is for an individual, or how their personal contribution might contribute to the wider society. It provides no interpretive framework through which policy may be guided. Finally, carbon footprinting uses no standard placement for the boundaries of the system in which it is deployed. Most calculations use 'cradle-to-gate', or 'cradle-to-site/plate' as the system boundary, while the least-used framework, and probably the most inclusive, is from 'cradle-to-grave'.

Another consumer-led policy initiative to close the 'carbon loop', and one triggered by the inter-governmental agreements at the first Earth Summit in 1992, which heralded the Clean Development Mechanism (CDM), is the development of voluntary carbon offsets. Carbon offsetting was seen as an approach with considerable appeal to environmentally conscious consumers, which might help assuage the guilt of people who travelled frequently by plane, but were painfully aware of the carbon cost of doing so. Offsetting flights is widely promoted as a solution to emissions reduction. It involves travellers paying a fee on top of their airfare to 'offset' the carbon emitted by the journey. However, there is considerable confusion surrounding carbon offsets: the way that emissions are measured, the fees charged for managing offsets and the methods employed in calculating them are all contentious and complex calculations (Gossling 2000). In addition, the main target of voluntary offsets has been tourists rather than the more significant business traveller, for whom there is evidently less appeal in 'guilt-free flying' (Francis 2009). The operator Responsible Travel, which has pioneered ethical tourism in Europe, has recently dropped its offsetting choices, on the grounds that some tourists might travel *more* because they believe the effect of their flights has been neutralised. Critics of offsetting argue that it has a negligible effect on carbon sinks in the global South, and, indeed, removes the responsibility for preventing deforestation in the developing countries themselves (Draper *et al.* 2009; Dawson *et al.* 2010).

Finally, in all the discussion of carbon accounting, trading and offsetting, there is a beggar at the feast. What might happen when markets fall and the price of carbon drops significantly? This eventuality had not received much attention in the optimistic decade that preceded the economic recession. In addition, some commentators have argued that in the European Union we are now faced by a 'sub-prime' market in carbon as the price drops, and investors lose the benefits of government support. This is a situation not entirely dissimilar to that in the housing market a decade earlier.

The shift towards more conventional policy tools, especially regulation, may also have political consequences, as the environmental movement in all its complexity assumes the lobbying role that has been the specialism of business and the environment since the ascent of ecological modernisation.

Underlying structural issues

On closer examination the sociological consequences of these developments in carbon markets are profound. The pricing of individual household-level technology, like wind turbines and solar panels, makes some Green innovations look more like 'positional goods' than public goods: the gains accruing mainly to people who can put up the initial capital, utilise government subsidies and, ultimately, reap the income benefits (Hirsch 1977). The contribution of these consumer 'fixes' to national energy production and the emerging 'energy gap' in the second decade of the century is likely to be very modest indeed. The observation that has been made about other forms of Green or ethical consumerism, such as 'responsible travel' and 'fair trade', also applies to these new forms of Green consumerism too: they require a higher outlay and yield most benefits for the middle-class consumer. Meanwhile, the big decisions on energy generation and conservation are stalled, and the nuclear power 'option' resurrected.

The emergence of Green technology as a positional good is only one of several indications that climate policy is insufficiently grounded in our knowledge of social structures. The existence of embodied carbon, and its acknowledgement in the discussions (but not the policies) surrounding global trade agreements, is another (Kejun and Murphy 2008). Climate policy, and the piecemeal attempt to provide incentives for individuals to reduce their own carbon dependency, is rarely linked to wider global experience outside the OECD countries. In what ways does it contribute to the transfer of much-needed cleaner technology to the global South? What are the international and distributive consequences *within* the global South of our attempts at limited decarbonisation in the North?

We might, indeed, dig deeper still. What other forms of human agency, other than those of the 'informed' consumer, have been left out of the equation? What are the wider social and cultural implications of placing so much emphasis on trading in a 'bad' (pollution) rather than a 'good' (such as cleaner technology)? What forms of human agency, innovation and collective action lie outside the compass of 'entrepreneurship', but help distil community support and engage environmental citizens (Dobson 2003)? Climate scientists are seen as the 'guardians of the dogma' on climate change, but there is evidence of low levels of public trust in science, including climate science. What is required, then, to mobilise areas in which there are high levels of public mistrust, such as climate change, while other institutions and practices do command widespread public support, such as community-based credit unions and some of the financial mutuals? New forms of Web communication and networking suggest widespread support for organisations which are embedded locally in communities and which acknowledge, rather than ignore, social and economic inequalities. As in previous historical periods, addressing structural inequalities, international as well as national, may become the engine of new transitions, creating new social solidarities, and means of liberation, from the path dependency associated with our heavy reliance on hydrocarbons (Redclift 2008).

Is there a 'bright narrative?'

In this chapter I have argued that a meaningful transition to a low-carbon economy is impossible as long as we rely on models of market choice and normative science policy that leave little room for collective and group behaviour and ignore the underlying social commitments that govern our everyday lives. The dual crises of global financial debt and climate change are reaching a 'tipping point' beyond which it will be difficult to move.

Already there is evidence that some behavioural responses to the environmental and financial crisis are taking forms that are not easily accommodated to the prevailing approaches to environmental policy favoured by most governments. They lie in challenges to conventional food systems, alternative recycling and reuse activities, small-scale attempts to provide sustainable renewable energy at the level of communities as well as individual households, and the brave efforts of enthusiasts to hold back ecological damage. Much of this activity is 'informal' in a new sense, too: it is often funded within the 'formal' market economy but depends heavily on Web-based organisation and group and community loyalties without formal institutional ties. These partial but evolving challenges to conventional thinking and behaviour are often only weakly connected to each other, since they cover a number of apparently isolated social fields. What they do reveal are fissures in the fabric of governance and the management of nature, and a need felt by some third-sector organisations to transcend anxiety over the environment. They reveal ways in which conventional path dependency is shifting, allowing new kinds of social organisation and governance to emerge, often in unexpected places, building new forms of social and ecological resilience. Can alliances be built from these small innovative 'alternatives'? Can a 'brighter narrative' be developed for the future?

These fissures also reflect a more profound underlying problem which makes responding to climate change more difficult. This is that many grassroots 'alternatives' are linked to oppositional groups opposed to 'globalisation', and indeed, in many cases originated from within these movements. The capacity to construct alternative futures, many of them utopian, has been a characteristic of Green and Left thinking, and has a long intellectual history (Kumar 1995). At the same time social movements provide examples which have a momentum of their own, often looking outside the consensus of liberal democratic institutions. Occasionally they catch the *Zeitgeist* for a moment; as during the early stages of the Obama campaign for President of the United States. Although globalisation, like climate change, is not something we can stand 'outside', it has some characteristics that distinguish it from the design of low carbon futures. First, of course, the technologies and organisation of low carbon futures can itself be assisted, or even driven, by the economic engine of globalisation. Second, unlike the antglobalisation movement, the advocates of reduced dependence on hydrocarbons are neither necessarily radical nor Green. The nuclear lobby is a case in point.

Societies and economies have been mobilised for different purposes at different times. Experiences in developing a critical stance towards globalisation can help us to bridge the gap between intellectual critique and alternative social practices. In exploring the possibilities of transition to a post-carbon future we might begin by examining the 'pieces' – fragmented, virtual and local – with which such a 'bright' narrative might be constructed. It will need to be constructed from people's lives and the resilience of their households and communities, rather than simply from their performance in consumer markets that are often transitory and unstable. But this is unlikely to be enough to redirect economic development in ways that are genuinely sustainable. The 'bright narrative' will elude this generation as long as the move to reduce carbon dependency remains disconnected from the struggles over democratic accountability, which remains the *sine qua non* for delivering sustainability in the next generation.

Note

1 There have been several reports suggesting that the European Union's Emissions Trading Scheme (ETS) will do little to encourage investment to reduce emissions during the economic recession. On the present course emissions trading is likely to produce only a 3 per cent reduction in emissions within the EU by 2020. Two effects will be observed. First, the cap on emissions will exceed projected EU emissions providing no economic incentive to move to clean technology and infrastructure before 2012. Second, because the EU allows unused permits and offsets under Phase 3 (2013–2020), any claimed economic incentive during this later period will be reduced also ('Recession plus ETS = fewer carbon emissions in the EU', National Audit Office Report, March 2009).

References

Adams, W.M. (1990) *Green Development*, London: Routledge.

Allan, J.A. (2003) Virtual water – the water, food and trade nexus: useful concept or misleading metaphor? *Water International* 28: 4–11.

Becken, S. (2005) Harmonising climate change adaptation and mitigation: the case of tourist resorts in Fiji. *Global Environmental Change* 15(4): 381–393.

Bellamy-Foster, J. (2010) Marx's ecology and its historical significance. In Redclift, M.R. and Woodgate, G. (eds) *The International Handbook of Environmental Sociology* (2nd edn), Cheltenham: Edward Elgar.

Brandt Commission (1980) *North–South: A Programme for Survival*, London: Pan Books.

Calder, L. (1999) *Financing the American Dream: A Cultural History of Consumer Credit*, Princeton, NJ: Princeton University Press.

Dawson, S.E.J., Lemelin, H. and Scott, D. (2010) The carbon cost of polar bear viewing tourism in Churchill, Canada. *Journal of Sustainable Tourism* 18(3): 319–336.

DEFRA (2008) (Department for Environment, Rural Affairs and Agriculture), Whitehall, London.

Demeritt, D. (1998) Science, social constructivism and nature. In Braun, B. and Castree, N. (eds) *Remaking Reality: Nature at the Millennium*, London: Routledge.

Dobson, A. (2003) *Citizenship and the Environment*, Oxford: Oxford University Press.

Draper, S., Goodman, J., Hardyment, R. and Murray, V. (2009) *Tourism 2023: Four Scenarios, A Vision and a Strategy for UK Outbound Travel and Tourism*, Forum for the Future, October.

Durning, A. (1992) *How Much Is Enough? The Consumer Society and the Future of the Earth*, New York: Norton.

Francis, J. (2009) Responsible travel has ditched offsetting flights and holidays for environmental reasons. *Guardian (London)*, 16 October.

Gossling, S. (2000) Sustainable tourism development in developing countries: some aspects of energy use. *Journal of Sustainable Tourism* 8(5): 410–425.

Guidolin, M. and La Jeunesse, E. (2007) The decline in the US. Personal savings rate: is it real and is it a puzzle? *Federal Reserve Bank of St. Louis Review*, November/December.

Hirsch, F. (1977) *The Social Limits to Growth*, London: Routledge.

IPCC (Intergovernmental Panel on Climate Change) (2007) *Fourth Assessment Review*, Cambridge: Cambridge University Press.

Kejun. C.A. and Murphy, D. (2008) Embodied Carbon in Traded Goods. Trade and Climate Change Seminar, 18–20 June, Copenhagen, IISD.

Kumar, K. (1995) *From Post-Industrial to Post-Modern Society*, Oxford: Blackwell.

Lovins, A., Hawken, P. and Hunter Lovins, L. (2000) *Natural Capitalism*, London: Earthscan.

Manning, R. (2001) *Credit Card Nation: The Consequences of America's Addiction to Credit*, New York: Basic Books.

Meadows, D.H., Meadows, D.L., Randers, D. and Behrens, F. (1972) *The Limits to Growth*, London: Pan Books.

Norgaard, R. (1988) Sustainable development: a co-evolutionary view. *Futures* 20(6): 606–620.

OECD (Organisation for Economic Cooperation and Development) (2004) *Measuring Sustainable Development: Integrated Economic, Environmental and Social Frameworks*, Paris: OECD.

Parker, J. (1999) Spendthrift in America? On two decades of decline in the US savings rate. *NBER Macroeconomics Annual* 14(1): 305–370.

Paterson, M. (2008) Global governance for sustainable capitalism? The political economy of global environmental governance. In Adger, N. and Jordan, A. (eds) *Governing Sustainability*, Cambridge: Cambridge University Press.

Pearce, D. (1991) *Blueprint Two*, London: Earthscan.

Princen, T., Maniates, M. and Conca, K. (eds) (2002) *Confronting Consumption*, Cambridge, MA: MIT Press.

Redclift, M.R. (1987) *Sustainable Development: Exploring the Contradictions*, London: Routledge.

Redclift, M.R. (1996) *Wasted: Counting the Costs of Global Consumption*, London: Earthscan.

Redclift, M.R. (2008) The Environment and carbon dependence: landscapes of sustainability and materiality. *Current Sociology* 57: 369–387.

Redclift, M.R. (2010) The transition out of carbon dependence: the crises of environment and markets In Redclift, M.R. and Woodgate, G. (eds) *The International Handbook of Environmental Sociology* (2nd edn), Cheltenham: Edward Elgar.

Redclift, M.R. and Hinton, E. (2008) Progressive Governance Conference: Living Sustainably. London: Policy Network.

Sachs, W. (2010) Globalisation, convergence and the Euro-Atlantic development model. In Redclift, M. and Woodgate, G. (eds) *The International Handbook of Environmental Sociology* (2nd edn), Cheltenham: Edward Elgar.

Schmidt, H. (2009) Carbon foot-printing, labelling and life cycle assessment. *International Journal of Life Cycle Assessment, Supplement 1*, S6–S9.

Simms, A. (2005) *Ecological Debt: The Health of the Planet and the Wealth of Nations*, London: Pluto Press.

Smith, N. (2007) Nature as accumulation strategy. In Panitch, L. and Leys, C. (eds) *Coming to Terms With Nature*, Socialist Register.

Soper, K., Thomas, L. and Ryle, M. (eds) (2008) *Counter-Consumerism and its Pleasures*, Basingstoke: Palgrave.

Stern, N. (2007) *The Economics of Climate Change: The Stern Review*, Cambridge: Cambridge University Press.

Stiglitz, J. and Serra, N. (2008) *The Washington Consensus Reconsidered*, Oxford: Oxford University Press.

Swyngedouw, E. (2009) The antinomies of the post political city: in search of a democratic politics of environmental production. *International Journal of Urban and Regional Research* 33(3): 601–620.

Welford, R. (2008) The economics of climate change: an overview of The Stern Review. *International Journal of Innovation and Sustainable Development* 1(3): 260–262

Woodgate, G. (2010) Introduction. In Redclift and Woodgate (eds).

Yearley, S. (1996) Rethinking the Global. In *Sociology, Environmentalism, Globalisation: Rethinking the Globe*, London: Sage.

3 Policy discourses of resilience

Katrina Brown

Introduction: how resilience has become top of the policy agenda

Resilience is currently infusing the agenda in climate change and international development as in many other areas of policy, and is promoted by different governments, non-governmental organisations and think-tanks. Resilience clearly has policy traction, but it is also not without its critics, who highlight in particular the lack of consideration of agency and power in the way resilience is promoted as a normative concept. In many respects then, the concept is rapidly gaining salience in environmental and development policy and in research at the interface of natural and social sciences, despite a lack of rigorous theoretical and empirical grounding in social sciences. This chapter analyses these policy discourses of resilience and climate change within the context of international development.

Resilience is a term in common usage but one which has been widely adopted in policy discourses over the past few years. Resilience ideas are evident in public discourse in a wide range of areas: economics, politics, environmental change, community development as well as health, well-being and poverty. Resilience is a characteristic or set of characteristics, applied variously to individuals, to families and communities, to cities, nations and to different systems – ecological, engineering and economic. Figure 3.1 shows how the use of resilience has increased over the past six years in terms of news (lower line) and reference searches (upper line) as reported by Google Trends.

There are three areas where resilience is especially prominent in both public and policy discussions which intersect and have particular relevance to political responses to climate change; these concern national security, disasters, and environmental change more broadly.

Resilience is a core topic in discussions of national, international and human security. Cascio (2009) suggested resilience as 'the next big thing' in *Foreign Policy* magazine's review of emerging trends, highlighting the inability of security to encompass profound uncertainty, non-linearities and surprise in social systems (see also Evans *et al.*, 2010). Resilience is increasingly linked to security where building resilience is promoted as a means to enhance security, both to reduce threats and their impacts, and to recover from disturbances. Examples

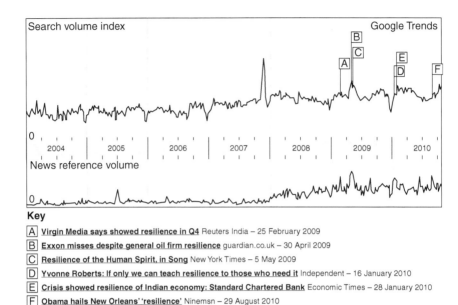

Key

A **Virgin Media says showed resilience in Q4** Reuters India – 25 February 2009

B **Exxon misses despite general oil firm resilience** guardian.co.uk – 30 April 2009

C **Resilience of the Human Spirit, in Song** New York Times – 5 May 2009

D **Yvonne Roberts: If only we can teach resilience to those who need it** Independent – 16 January 2010

E **Crisis showed resilience of Indian economy: Standard Chartered Bank** Economic Times – 28 January 2010

F **Obama hails New Orleans' 'resilience'** Ninemsn – 29 August 2010

Figure 3.1 Resilience in the news and on the World Wide Web.

include national, regional and city-scale initiatives such as 'Scottish Resilience', 'Birmingham Resilience' and 'Resilient US'. The UK's National Security Strategy (Cabinet Office, 2008), for example, specifies resilience as a goal of national security. The Strategy highlights resilience of communities both as a goal and as a means to overall security. But the Security Strategy focuses heavily on resilience in the area of civil contingencies rather than resilience as a cross-scale system property (see also IPPR Commission on National Security, 2008).

The term resilience is increasingly used in relation to disasters and how individuals, communities and nations are able to respond to shocks and contingencies. For example, in response to the 2010 floods in Pakistan, Patrick McCully, writing in the *Huffington Post*, argues that the catastrophe demonstrates that fundamental systematic change is required in river basin management ('Increasing resilience to floods in Pakistan, the US, and just about everywhere else is going to require reversing our river management mistakes through restoring rivers and floodplains, including by taking out embankments and dams'; see McCully, 2010). The World Food Programme – along with other commentators and NGOS – highlights resilience of individuals, families and communities to multiple stressors in linking flood with conflict (see WFP, 2010). A strong thread in this field is that 'ordinary' people exhibit resilience in the face of these events in spite of government and other policies which have provoked events or made people and places more vulnerable. Resilience in this context is the ability to withstand and cope, and to recover from an event.

The adoption of resilience ideas and slogans in relation to environmental change resonates with these areas too. Resilience ideas are evident in international science and policy statements such as the Millennium Ecosystem Assessment (2005); in the *Human Development Reports* from UNDP (UNDP, 2007); the Swedish government-sponsored Commission on Climate Change and Development (2009); and initiatives such as the World Bank's *Program for Climate Resilience* (World Bank, 2008). Non-governmental organisations including Christian Aid and Oxfam, and think-tanks such as the World Resources Institute have used the term to frame their policy documents and analysis (Brown, 2010). Terms such as climate change resilience and climate resilient development have become prevalent. These are discussed in more detail below. Before analysing how the concept of resilience is being used in policy discourses on climate change, this chapter provides a brief overview of conceptualisations of resilience in different areas of natural and social sciences to understand its different meanings and multiple dimensions.

Framing resilience: resilience thinking across disciplines

Resilience ideas emerge from a number of different fields. Norris *et al.* (2008) trace the origins of resilience to physics and mathematics, where the term was used to describe the ability of a material or system to return to equilibrium following a displacement – a resilient material bends or bounces back rather than breaks when stressed, for example. This indeed reflects its general understanding and usage. The term has since been applied to describe the adaptive capacities of individuals, human communities and larger societies, as well as ecosystems, cities, economies, nations, and particularly latterly social ecological systems. Norris *et al.* (2008: 129) present 21 representative definitions of resilience from diverse disciplines including psychology, sociology, geography, anthropology, public health, ecology, technology and communications. In these definitions resilience is applied to physical systems, ecological and social systems, cities, communities and individuals.

Box 3.1 shows definitions of resilience from three key fields which inform its current applications in climate change. The first comes from the literature on social-ecological systems – which emerges from ecology, extended to encompass social systems and institutions, and promoted especially through the Resilience Alliance, an international scientific network of scientists interested in the analysis of resilience. This definition takes an explicit systems approach. The second definition, from Michael Rutter (2004), comes from child developmental psychology where resilience represents the capacity of an individual to successfully adapt in the face of adversity. The third definition, relating to health and disasters, is slightly broader in that it applies to individuals, families, communities and systems to deal with shocks (Almedom, 2008). Yet despite these diverse fields there are a number of key similarities, and each of these literatures uses resilience as describing a state and as a set of characteristics, capacities or processes which confer adaptability or adaptedness (see Nelson *et al.*, 2007) and this helps to broadly characterise a resilience approach.

A number of critical issues are common to understandings of resilience across these disciplines. They each view change as part of how a system works, and so change is expected and management strategies must be based on this assumption. Furthermore, uncertainty and surprise are also features; not all change is predictable nor is it linear or gradual – it may be uneven, and may be characterised by thresholds or 'tipping points' both social and ecological. There are different types of change, sometimes classified as fast and slow variables, and there are interactions and linkages – and feedbacks – between different stressors. The responses to change may also be distinguished as coping, adapting and transforming. Each of the disciplines – or areas of study identified above – also recognises that change and even shocks or disturbances are not always detrimental and that crises may provide 'windows of opportunity', a stimulus to shift from established patterns to more (perhaps) beneficial ones, which may change the distribution of costs and benefits between different individuals or sections of society. Finally, each of these approaches also sees resilience as a property of systems at different scales, and recognises the significance of cross-scale interaction – for example, in the shape of 'panarchy' in the social ecological systems literature where resilience and change at specific scales can influence or control dynamics at others (Gunderson and Holling, 2002), and individual, family and community in the human development literature (Masden and Obradovic, 2008; Norris *et al.*, 2008). Thus, in each of these perspectives, resilience is a dynamic property.

In reviewing 16 conceptualisations of resilience across literature related to environmental change, development and disasters, Bahadur *et al.* (2010) identify ten common characteristics of resilience. These include properties of the system itself, such as high diversity, uncertainty and change, non-equilibrium dynamics; and its related institutions, such as effective governance, community participa-

---

Box 3.1 Definitions of resilience

Resilience definitions

> The capacity of a system to absorb disturbance and reorganize while undergoing change so as to still retain essentially the same function, structure, identity, and feedbacks.
>
> (The RA website glossary at www.resalliance.org/ after Walker *et al.*, 2004)

> The process of, capacity for, or outcome of successful adaptation despite challenging or threatening circumstances
>
> (Rutter, 2004)

> A multi-dimensional construct ... the capacity of individuals, families, communities, systems and institutions to respond, withstand and/or judiciously engage with *catastrophic* events and experiences; actively making meaning without fundamental loss of identity.
>
> Source: Almedom (2008)

tion, equity, learning, preparedness, and then cross-scale interactions and supportive social values and structures. As in reviews from other fields (e.g. Ungar *et al.*, 2007), Bahadur *et al.* highlight the tensions between understandings of resilience as an *outcome* and as a *process*. This resonates with further observations about the overlaps and confusion between resilience and resistance (e.g. Jerneck and Olsson, 2008). Bahadur *et al.* identify four problems with the conceptualisation of resilience in the environmental change, development and disasters fields (2010: 19). First, there is a lack of clarity about the relationship between adaptation, adaptive capacity and resilience; second, the boundaries and limits of the 'system' to which resilience applies are rarely defined; third, there is a major gap in how resilience can be measured; fourth, most discussions are conceptual, and there is a lack of robust case studies to test theories. These issues clearly have implications for how resilience is applied in different areas of policy and how resilience approaches can be implemented and evaluated.

But there are a number of criticisms of resilience ideas, their assumptions and prescriptions, and how they are applied in the field of environmental change which require interrogation from social scientists in particular, and these have implications for theory and conceptualisation, as well as for policy and practice (see e.g. Nadasdy, 2007; Leach, 2008; Hornborg, 2010). They view resilience thinking as depoliticised, and lacking in appreciation of agency. Fundamental to these is the mistrust by many social scientists of a concept which is based so overtly on a systems perspective and which is therefore seen, among other things, to under-represent issues of power (Nelson *et al.*, 2007). Of particular relevance is the tension between resilience as an inherently dynamic concept and its emphasis on maintaining structure, function and identity. At the heart of a resilience approach is the recognition of multiple stable states or equilibria – that is why it is such a paradigm-shifting concept in ecology – but what use is resilience if you want to change structures? This is especially pertinent in debates around climate change adaptation – in considering equity, for example – and mitigation, in contemplating transition to a 'low-carbon society'.

Policy discourses on resilience and climate change

As noted, resilience ideas are especially prevalent in discussions on climate change and are being widely promoted in a number of responses. Resilience is a concept which is being widely applied in policy on climate change and international development. A number of international agencies (including UNDP, World Bank, European Commission) and the UK DFID promote resilience as a means to link development efforts to climate change. Table 3.1 provides an overview of a few of the policy prescriptions which relate to climate change in the context of international development.

Within these policy discourses, resilience is applied to a number of different entities. *Ecological*, *social* and *economic resilience* are identified and resilience is seen as a property of communities (Christian Aid, n.d.; WRI, 2008), the rural poor (WRI, 2008), countries and states (World Bank, 2009a), as well as of

Table 3.1 Policy prescriptions on resilience and climate change

Reference	How resilience is defined	Prescriptions
WRI (2008) Roots of Resilience	Resilience is the capacity of a system to tolerate shocks or disturbances and recover – applied specifically to communities.	Resilience means communities are better prepared to survive economic downturn, environmental challenges and social disruptions; based on the need to scale up the resilience of the rural poor to accommodate social and environmental change, particularly that which is arising from climate change.
World Bank (2008) Pilot Program for Climate Resilience under the Strategic Climate Fund	Resilience not specifically defined, but aims to 'integrate climate risk and resilience into core development planning, while complementing other ongoing activities'.	Aim to provide incentives for scaled-up action and transformational change in integrating consideration of climate resilience in national development planning consistent with poverty reduction and sustainable development goals. (section B paragraph 4).
Christian Aid (n.d.) Overexposed: Building disaster-resilient communities in a changing climate	Resilience is of ability to withstand the impact of shocks and crises. It is determined by people's assets and their ability to access services provided by external infrastructure and institutions.	The Building Disaster-resilient Communities project supports local partner organisations … to strengthen communities' abilities to manage and recover from crises and to prepare for and reduce the risk of future disasters (p. 8).
Oxfam (2009) People-centred Resilience	Resilience is the ability of a joint social and ecological system – such as a farm – to withstand shocks, coupled with the capacity to learn from them and evolve in response to changing conditions. Building resilience involves creating strength, flexibility and adaptability.	Makes the case for investing in building up the resilience of vulnerable farming communities as a critical stepping stone to addressing the global challenges of food security; climate change adaptation; and climate change mitigation (p. 7).

infrastructure and even enterprise. The threats to resilience are various, but climate change is identified in all as a common threat, but is linked to other types of change in different ways. For example, the WRI (2008) defines resilience of the rural poor to environmental and social challenges including climate change, loss of traditional livelihoods, political marginalisation, and breakdown of customary village institutions. Most of these policy documents emphasise climate change but reflect in particular the context of poverty and impoverishment in developing countries, and also increased risks and precarious livelihoods, and ongoing environmental degradation.

The prescriptions offered in these documents are quite different, but they do have important common features. The WRI's view of building resilience involves 'scaling up' – enhancing local-scale projects and funding – and also necessitates drawing people into markets, especially markets for ecosystems services as a means to deliver what they refer to as the 'Resilience Dividend' to the rural poor. They propose this as a 'community driven model to manage their ecosystem assets and build them into enterprises that can experience a marked increase in their resilience' (WRI, 2008: 27). The core idea is that markets and particularly payments for ecosystem services provide positive case studies of the 'power of self-interest and community ownership, the enabling value of intermediary organisations and how communication and networks can provide new ideas and support' (WRI 2008: 3). A number of different metaphors and narratives are employed to explain the dynamics of change and to justify the idea of nurturing ecosystem-based enterprises, using terms such as 'Green Livelihoods' and 'Turning back the Desert'. The enterprise model presented includes both adaptation and mitigation of climate change (for example, drawing smallholders into carbon markets), and presents overall the potential of positive outcomes and opportunities from finding innovative responses to change. In this respect it relates strongly to prescription from other agencies (e.g. FAO, 2008) which promote payments for ecosystem services as a means of alleviating rural poverty and addressing environmental concerns.

In contrast to the WRI's rather optimistic view of resilience as enabling and building capacity, the Commission on Climate Change and Development (2009) uses resilience in a quite different way and presents a different set of prescriptions. The Commission views resilience as much more concerned with managing and reducing risks. It has a specific focus on the poorest, and presents resilience in terms of capacities, closely aligned with disaster risk reduction, and the need for greater knowledge and communication. It stresses the need to build the resilience of communities, countries and regions to cope with unexpected events. The Commission quotes Patrick Lagadec, and emphasises the aim 'not to strive to foresee the unforeseeable but to train ourselves to cope with it … not to clarify, map and plan for every single surprise, but to train to be surprised' (Commission on Climate Change and Development, 2009: 77).

Disaster risk reduction has taken on new meanings and has been given renewed momentum for many development agencies in the face of climate change. Christian Aid (n.d.) emphatically states that climate change will increase the number and impact of various types of disasters around the world. The emphasis of its policy, exemplified in the Building Disaster Resilient Communities Project, is on disaster risk reduction; integrating early warning systems and linking local, subnational and national systems. It prioritises the need to 'climate proof' development and humanitarian work. Again though, it is difficult to see what is novel in the resilience approach; for example, it stresses the need for strengthening water and watershed management in response to 'new' dangers from climate change, manifest in slow onset disasters such as drought.

Some of these themes resonate with the rhetoric and policies of the World Bank, through its 2010 World Development Report (World Bank 2009a), and specifically its objective of climate resilient growth. This aims to address climate change risks within project design, and to integrate risk and build capacity in developing countries. It again links adaptation and mitigation, and sees integration into markets as a key means of achieving this. Thus resilience is used – as in the WRI document discussed above – to promote existing agendas concerning expanding markets, strategies which may generate new vulnerabilities and inequalities. Raising agricultural productivity, building knowledge infrastructure, providing energy for all, are also core elements. The *Pilot Program for Climate Resilience (PPCR) Under the Strategic Climate Fund* (World Bank, 2008) is a major means of implementing these objectives, with specific regional strategies developed (e.g. the strategy for Africa; World Bank, 2009b). As discussed in detail in the following section, the emphasis here is on making current development activities resilient to climate change, but the main focus is still on growth, productivity and markets.

There are multiple and conflicting discourses in each of the policy statements, and importantly there is inconsistency in how terms are used and ideas applied. Many of the statements do not capture the complex and dynamic systems approach that resilience thinking encompasses. For example, the system or the basic entity is poorly defined – only the WRI document links the environmental and social in any 'systems' conceptualisation or framework (this mentions a *Social Ecological System*). Most of the other documents refer to a poorly defined entity. Other tenets of resilience thinking, including thresholds, feedbacks; networks, connections; transformation, transformative change, are almost completely absent. The areas where there is stronger correlation to the conceptual or theoretical literature are in some of the approaches to promoting resilience through applying adaptive management and in disaster risk reduction. But the problems identified by Bahadur *et al.* (2010) are clearly evident, and especially the tendency to promote resilience in order to maintain some form of stability – as WRI articulates: '*increased resilience results in ecosystem stability, social cohesion and adaptability, economic enterprise*' (WRI, 2008: 6), and the need to scale up the resilience of the poor to *accommodate* environmental and social change which is a passive notion of responding to change. For Christian Aid (n.d.: 9), resilience is 'the ability to *withstand* the impact of shocks and crisis'. The emphasis on integration into the economic mainstream, into cash economies and markets is quite striking. The emphasis by the World Bank of making growth and current development resilient in the face of climate change is especially telling. The chapter now analyses this particular policy in more depth.

Unpicking 'climate resilient development'

Most prominent among the proponents of climate resilient development as a policy discourse and a set of implementable actions is the World Bank which has developed a programme to pilot and demonstrate ways to integrate climate risk and resilience into core development planning while supplementing and bolstering its ongoing activities. The Pilot Program for Climate Resilience (PPCR) is part of the Strategic Climate Fund (SCF), a multi-donor Trust Fund within the Climate Investment Funds. The overall objective of the programme is to provide incentives for scaled-up action and transformational change in integrating consideration of climate resilience in national development planning consistent with poverty reduction and sustainable development goals. Approved in 2008, the projected 'pledged resource envelope' has increased from US$614 million (September 2009) to US$975 million (July 2010) of which US$9 million has been disbursed to date (Climate Funds Update, 2010). This has been pledged by eight country donors: Australia, Canada, Denmark, Japan, Germany, Norway, the UK and USA. There are nine countries and two regions which have been invited to participate; the countries include Zambia, Mozambique, Bolivia and Bangladesh, and the regions are the Pacific and the Caribbean. Participation is contingent upon recipient countries fulfilling the criteria of the respective trust funds; that is, adopting bank and donor conditions in exchange for financing. For the PPCR, eligible countries will have to submit 'country investment strategies' which will be assessed by the SCF PPCR Subcommittee. Priority is given to highly vulnerable least developed countries eligible for MDB concessional funds, including the small island developing states.

In the PPCR it is development itself – the process of wealth generation – which is being made more resilient to the impacts of climate change. The PPCR is designed to provide finance for climate resilient national development plans and to build on existing National Adaptation Programs of Action (NAPAs). It provides only short-term funding and the purpose is to offer lessons over the next few years that may be taken up by countries, the development community, and the future climate change regime, including the Adaptation Fund.

Its objectives are summarised (Climate Funds Update, 2010) as:

- Pilot and demonstrate approaches for integration of climate risk and resilience into development policies and planning;
- Strengthen capacities at the national levels to integrate climate resilience into development planning;
- Scale up and leverage climate resilient investment, building on other ongoing initiatives;
- Enable learning-by-doing and sharing of lessons at country, regional and global levels.

The World Bank's strategy for climate resilient development in Africa (World Bank, 2009b) spells out what it means by climate resilient development in more detail. It centres on making adaptation a core component of development, with a particular focus on sustainable water resources, land and forest, integrated coastal zone management, increased agricultural productivity, health problems, and conflict and migration issues. It claims to focus on knowledge and capacity development by improving weather forecasting, water resources monitoring, land use information, improving disaster preparedness, investing in appropriate technology development, and strengthening capacity for planning and coordination, participation and consultation. It emphasises the benefits from mitigation opportunities through access to carbon finance against land use changes and avoided deforestation, promoting clean energy sources (e.g. hydropower) and energy efficiency, and adopting costeffective clean coal energy generation and reduced gas flaring. It also highlights the opportunities of 'scaling up financing', in other words maximising flows of these conditional investments. Thus it may be seen to reinforce dominant development agendas based on economic growth. But these strategies themselves may have inherent biases – benefiting some sections of society over others – and may reinforce or bring yet new vulnerabilities.

Clearly these suggested climate change and development policies are *not* radical – they echo market environmentalism, ecological modernisation and environmental populism. The proposals encourage an incremental rather than a radical approach. The discourse here sees resilience as part of a strategy which mainstreams climate change adaptation and mitigation into development efforts. It sees climate change as a challenge to current development but also as providing opportunities – especially for attracting investments (via carbon markets and market-based mitigation especially in the case of Africa in land use, forest management and renewable energy). It suggests that adaptation is 'fundamentally about sound, resilient development'; resilience in this context is bestowed by mainstreaming climate change; by making adaptation and risk management core development elements; by fostering knowledge and capacity (particularly in terms of access to information and forecasting); and scaling up financing. There are explicit links made to disaster preparedness and disaster risk reduction, to providing layers of insurance protection, and safety nets where appropriate, and to building 'climate-smart' systems.

In this way the approach mixes a number of different aspects of resilience thinking, including multiple and cross-scale dynamics; the emphasis on shocks and disturbances to the system; but also aspects of hard, engineering-like resilience. Importantly, climate resilient development is not presented as anything fundamentally different to current development – it emphasises that current plans need to be 'climate proofed'; in other words, the potential future impacts of climate change and associated risks must be built in. But the premise of continued growth and the benefits of this strategy are not questioned; climate change re-enforces the need to do development 'better', more effectively, and with an emphasis on shifting vulnerabilities and how they may reconfigure the distribution of costs and benefits within society.

This mirrors discussions on climate change adaptation, where distinctions can be made between approaches which see adaptation as necessary to protect development, and those which see climate change and adaptation as an opportunity to change development (see e.g. Ensor and Berger, 2009).

Resilience as business-as-usual

Current uses of resilience in international development policy as a response to climate change promote business as usual; quite contrary to the emphasis on change in 'resilience thinking' they emphasise a defence of the status quo. In each of these policy approaches and discourses resilience is being applied in a normative sense, and overwhelmingly as a means to protect the status quo; to resist or accommodate change; and to enhance stability rather than dynamic responses. The prescriptions are overwhelmingly technocratic and managerialist, which Brooks *et al.* (2009) suggest is inevitable given the dominant modernist developmentalist paradigm. McMichael identifies climate proofing as a 'new profit frontier' (2010: 252), emphasising how the response of multilateral agencies to climate change is to 'marketise' development adaptation.

But is there scope for a more radical interpretation and application of resilience ideas to inform international development in the light of climate change? Can resilience ideas support these types of transformation? At a conceptual level, resilience as a system property is wholly compatible with making choices about the type of change and about dealing with different types of change in a more proactive rather than a merely reactive sense. But resilience has a dark side. This is evidenced through the analysis of policy statements presented here, and in how resilience ideas are being promoted in other fora. These issues were rehearsed at a recent meeting on resilience and indigenous people, where resilience concepts were discussed and contested. The debate is summarised in a paper based on the discussions at the meeting (Rotarangi and Russell, 2009: 211) which claims that 'to be indigenous is to be resilient' and cautions that 'western scientific applications are in danger of being applied as yet another colonising model, retaining power and legitimised knowledge in the hands of the dominant society'. The multiple and even conflicting meanings of resilience are also apparent in how local communities challenge government policy in my own research exploring narratives of resilience in coastal communities in the UK (Brown and Hayward, 2008; Tompkins *et al.*, 2008). This shows how an important part of communities' perception of resilience is articulated as their capacity to resist externally driven policy (in this case coastal zone management in response to climate change) – for them, resilience means self-determination. Ironically, the policies justified on the grounds of scientific resilience – specifically those for managed retreat or inundation in the case of East Anglia – are those to which local communities are especially resistant. Thus issues of resilience are intimately intertwined with issues of power, knowledge, justice and self-determination.

Jerneck and Olsson (2008) analyse resilience as one of the dominant discourses surrounding adaptation to climate change. They identify the implicit normative assumption of the preservation of the system and thus relate resilience to resistance to change. Hence, the need for the ecological system to remain within its 'basin of attraction' (Walker *et al.*, 2004) has implications for the social system; they argue that there is a built-in contradiction in the concept of resilience when it is applied to complex systems where subsystems with conflicting goals are linked. They maintain that resilience may be useful for guiding adaptation, for instance, in systems where there is no inherent conflict between social and ecological components of systems. But they contend that a resilience approach (exemplified by insurance-based approaches as promoted by the World Bank) might delay a long-term solution to vulnerability by encouraging people to continue with a livelihood for which there is no sustainable basis. This contradicts the view of Nelson *et al.* (2007) who would classify this example as a vulnerability-not-resilience approach; a resilience approach would take into account multiple stressors and system-wide change over different spatial social and temporal scales. But Jerneck and Olsson argue that ideas of resilience underline recovery more than fundamental change (2008: 179–180) and are thus more likely to support incremental rather than profound change as in the case of the policy discourses analysed here. Cannon and Muller-Mahn (2010) also highlight some of the underlying conceptual contradictions in resilience approaches; they contend that it promotes a scientific and technical approach akin to 'imposed rationality' that is alien to the practice of ordinary people. Like Nadasdy (2007), they identify a critical issue in how resilience is depoliticised and does not take account of the institutions within which practices and management are embedded. They argue that these problems stem from its origins in systems thinking and in transferring a concept derived from the analysis of natural systems to social phenomena. Nadasdy (2007) provides an explanation of why, despite a rhetoric stressing management for resilience which might be at the expense of short-term stability, there are powerful interests to protect against such a dynamic or adaptive strategy. He argues that capitalist production demands a permanent degree of short-term stability and, so long as this system remains, the pressure to make management decisions based on the stability of one or two key resources will remain enormous (2007: 217). Nasdasdy reminds us that capitalism simply cannot be viewed as a set of social processes and relations that play themselves out on a neutral landscape; present-day social ecological systems are themselves the products of capitalist processes and social relations. For McMichael, what is required is an ontological break with the standard market episteme of the development project and its 'global ecology' which has the metabolic rift at its foundation (2010: 259). The decentred, decommodified and decarbonised alternative could still have resilience ideas at its heart, but not as they are currently espoused.

References

Almedom, A.M. (2008) Resilience research and policy/practice discourse in health, social, behavioral, and environmental sciences over the last ten years. *African Health Sciences* 8 (Suppl.): S5–S13.

Bahadur, A.V., Ibrahim, M. and Tanner, T. (2010) The resilience renaissance? Unpacking of resilience for tackling climate change and disasters. Strengthening Climate Resilience Discussion Paper 1, IDS, Sussex.

Brooks, N., Brown, K. and Grist, N. (2009) Development futures in the context of climate change: challenging the present and learning from the past. *Development Policy Review* 27 (6): 741–765.

Brown, K. (2010) Policy discourses on resilience. Paper presented at the Association of American Geographers Annual Meeting, Washngton, DC, April.

Brown, K. and Hayward, B. (2008) Nowhere far from the sea: exploring resilience in coastal communities. Paper presented at Stockholm Resilience Symposium.

Cabinet Office (2008) *UK National Security Strategy*. London: Cabinet Office.

Cannon, T. and Muller-Mahn, D. (2010) Vulnerability, resilience and development discourses in the context of climate change. *Natural Hazards.* 55: 621–635.

Cascio, J. (2009) The next big thing: resilience. *Foreign Policy*, April. Available online at www.foreignpolicy.com/2009/04/15.

Christian Aid (n.d.) *Overexposed: Building Disaster-resilient Communities in a Changing Climate*. London: Christian Aid.

Climate Funds Update (2010) www.climatefundsupdate.org/listing/pilot-program-for-climate-resilience (accessed November 2010).

Commission on Climate Change and Development (2009) *Closing the Gaps: Disaster Risk Reduction and Adaptation to Climate Change in Developing Countries*. Final Report. Stockholm, Sweden.

Ensor, J. and Berger, R. (2009) *Understanding Climate Change Adaptation: Lessons from Community-based Approaches*. Rugby, UK: Practical Action Publishing.

Evans, A., Jones, B. and Steve, D. (2010) *Confronting the Long Crisis of Globalization: Risk, Resilience and International Order. Managing Global Insecurity*, Brookings/CIC: New York University.

FAO (2008) *Paying Farmers for Environmental Services: The State of Food and Agriculture 2007*. Rome: FAO.

Google Trends (2010) ww.google.com/trends?q=resilience&ctab=0&geo=all&date=all (accessed November 2010).

Gunderson, L. and Holling, C.S. (eds) (2002) *Panarchy: Understanding Transformations in Human and Natural Systems*. Washington DC: Island Press.

Hornborg, A. (2010) Zero-sum world challenges in conceptualising environmental load displacement and ecologically unequal exchange in the world-system. *International Journal of Comparative Sociology.* 50 (3–4): 237–262.

IPPR Commission on National Security (2009) *Shared Responsibilities: A National Security Strategy for the UK*. Final Report. London: IPPR.

Jerneck, A. and Olsson, L. (2008) Adaptation and the poor: Development, resilience and transition. *Climate Policy.* 8: 170–182.

Leach, M. (ed.) (2008) Re-framing resilience: trans-disciplinarity, reflexivity and progressive sustainability – a symposium report. *STEPS Working Paper 13*. IDS, Sussex.

Masden, A.S. and Obradovic, J. (2008) Disaster preparation and recovery: lessons form research on resilience and human development. *Ecology and Society* 13 (1): 9. Available online at www.ecologyandsociety.org/vol. 13/iss1/art9.

McCully, P. (2010) Global lessons from the Pakistan flood catastrophe. *Huffington Post.* Available online at www.huffingtonpost.com/patrick-mccully/global-lessons-from-the-p_b_691928.html (accessed November 2010).

McMichael, P. (2010) Contemporary contradictions of the Global Development Project: geopolitics, global ecology and the 'development project'. *Third World Quarterly* 30 (1): 247–262.

Millennium Ecosystem Assessment (2005) *Ecosystem Services and Human Well-being.* Washington, DC: Island Press.

Nadasdy, P. (2007) Adaptive co-management and the gospel of resilience, in D. Armitage, F. Berkes and N. Doubleday (eds), *Adaptive Co-management: Collaboration, Learning and Multi-level Governance.* Canada: UBC Press, pp. 208–227.

Nelson, D., Adger, N. and Brown, K. (2007) Resilience and adaptation to climate change: linkages and a new agenda. *Annual Review of Environment and Resources* 32: 395–419.

Norris, F.H., Steves, S.P., Pfefferbaum, B., Wyche, K.F. and Pfefferbaum, R.L. (2008) Community resilience as a metaphor, theory, set of capacities and strategy for disaster readiness. *American Journal of Community Psychology* 41: 127–150.

Oxfam (2009) *People-centred Resilience.* Oxford: Oxfam UK.

Resilience Alliance (2010) Glossary at www.resalliance.org/608.php#R (accessed November 2010).

Rotarangi, S. and Russell, D. (2009) Social ecological resilience thinking: can indigenous culture guide environmental management? *Journal of Royal Society of New Zealand.* 39 (4): 209–213.

Rutter, M. (2004) The promotion of resilience in the face of adversity, in A. Clarke-Stewart and J. Dunn (eds), *Families Count: Effects on Child and Adolescent Development.* New York: Cambridge University Press.

Tompkins, E.L., Few, R. and Brown, K. (2008) Scenario-based stakeholder engagement: incorporating stakeholders preferences into coastal planning for climate change. *Journal of Environmental Management* 88: 1580–1592.

UNDP (2007) *Human Development Report 2007/2008 Fighting Climate Change: Human Solidarity in a Divided World.* New York: Palgrave Macmillan.

Ungar, M., Brown, M., Liebenberg, L., Othman, R., Kwong, W.M., Armstrong, M. and Gilgun, J. (2007) Unique pathways to resilience across cultures. *Adolescence* 42 (166): 287–310.

Walker, B., Holling, C.S., Carpenter, S.R. and Kinzig, A. (2004) Resilience, adaptability and transformability in social-ecological systems. *Ecology and Society* 9(2): 5. [online] URL: www.ecologyandsociety.org/vol. 9/iss2/art5/.

World Bank (2009a) *World Development Report 2010: Development and Climate Change.* Washington, DC: World Bank.

World Bank (2009b) *Africa's Development in a Changing Climate.* Washington, DC: World Bank.

World Bank (2008) *Pilot Program for Climate Resilience under the Strategic Climate Fund.* Washington, DC. World Bank.

World Food Programme (2010) *Conflict Then Floods – Pakistani Family Shows Resilience.* News item, 4 October. Available online at 2010 www.wfp.org/stories/first-conflict-now-floods-pakistan-family-shows-resilience.

World Resources Institute (2008) *Roots of Resilience: World Resources Report.* WRI.

4 Resilience and transformation

Mark Pelling

Introduction

The adaptation and mitigation demands of climate change at the international level have long been connected to overarching debates and negotiations on sustainable development. The early progress of the UNFCCC lies partly in the Rio Earth Summit, 1992, where sustainable development was popularised as a global policy agenda. As part of this event the first Framework Convention on Climate Change agreement was opened for signature. Old sticking points that have blocked international consensus on ecological and social aspects of sustainable development have also been rehearsed in the UNFCCC negotiations with relative levels of wealth and development legacies being invoked. Beyond international negotiation strategies, the temporal quality of climate change response also echoes sustainable development with adaptation and mitigation decisions today having justice implications for future as well as current generations (Adger *et al.*, 2009). This connection to debates on sustainable development reminds us that climate change and responses are but contemporary expressions of an underlying and ongoing crisis in environment–society relationships. Solutions to climate change associated risk require a re-evaluation of development visions and practice and outcomes, as well as a focus on more proximate drivers of change and risk.

The target of adaptation is social behaviour framed in each context by specific risks or hazards of interest. Climate change influences risk in at least three interacting and overlapping ways: long-term trends in mean temperatures and other climatic norms, including secondary effects such as sea-level rise; temporal shifts in seasonality that bring locally extreme and unusual conditions; and increased extremes in variability that can trigger natural disasters such as floods, hurricanes, fires, etc. Adapting to the local impacts of these changes requires sensitivity to local context as other risks and opportunities (social, economic and political as well as environmental) shape and limit human well-being, the functioning of socio-ecological systems, and what is possible and preferred in adaptation (Pelling and Wisner, 2008).

Work examining processes of adaptation has benefited from a number of typologies of adaptive action (Smit *et al.*, 2000; Smit and Wandel, 2006; Burton *et al.*, 2007). Early work framed adaptation in terms of regional or national scale agro-economic systems. For example, Krankina *et al.* (1997) refer to boreal forestry management strategies as a means of assisting forests to adapt. Here the system of interest was ecological and the management system an intervening variable between it and climate change. This complements well the scale of resolution available from climate modelling, but is less suited to exploring the social processes driving and limiting adaptive decision-making which is always locally constituted. Alternative approaches to adaptation science have emerged that start from the viewpoint of local actors at risk. These approaches do a better job at contextualising adaptation within development and explaining why people are unable or unwilling to take adaptive action. For example, in an analysis of two communities in Puerto Rico, López-Marrero and Yarnal (2010) found that concerns for health conditions, family well-being, economic factors and land tenure were given more priority by local actors than adaptation to climate change, despite their exposure to flooding and hurricanes. The importance of conceptualising and managing adaptation within the context of multiple risks, and of people's general well-being, is clear.

If adaptation is to move us towards a more sustainable development path then changes in values and associated governance regimes will need to be on the agenda, or may force themselves on as established institutions lose legitimacy in the face of climate change associated economic, ecological and disaster loss. Such systems present limits to adaptation that are emotional as much as they are political and show the importance of considering the socially constructed levels of risk associated with climate change (Adger *et al.*, 2009). This may produce some surprising outcomes. Kuhlicke and Kruse (2009), for example, show how local adaptive actions to reduce flood risk along the Elbe River, Germany, rely mainly on anticipation and assumptions about state support, the latter actually being seen to undermine local resilience. Adaptation is also limited by institutional failures. This is the principal reason for physical infrastructural failure – the proximate cause for many events. Institutions fail to enable adaptation when those at risk and managing risk are not able to learn critically, but rather are trapped in cycles of marginal improvements of existing behaviour; when those at risk and their advocates cannot hold risk managers to account; and when information and resources cannot be used effectively or equitably (Wisner *et al.*, 2004). Finally, limits arise from the speed of development and application of appropriate technological innovations. In South Asia, in the space of a generation cell phone technology has enabled mobile phones to spread from being the preserve of the wealthy of a ubiquitous feature of urban and rural life alike with knock-on benefits including providing early warning for disaster risk (Moench, 2007). These accounts indicate the complexity of identifying limits to adaptation and the great sociological and geographically variation to be expected.

The co-production of risk and development

The possibility that adaptation can open up space for incremental and radical adjustment, as well as stability, is presented in Table 4.1 as a distinction between resilience, transition and transformation. Resilience refers to a refinement of actions to improve performance without changing guiding assumptions or the questioning of established routines. This could include the application of resilient building practices or application of new seed varieties. Transition refers to incremental changes made through the assertion of pre-existing unclaimed rights. This might include a citizens' group claiming rights under existing legislation to lobby against a development that would undermine ecological integrity and local adaptive capacity. Transition implies a reflection on development goals and how problems are framed (priorities, include new aspects, change boundaries of system analysis), and assumptions about how goals can be achieved. Transformation refers to irreversible regime change. It builds on the recognition that paradigms and structural constraints impede widespread and deep social reform (for example, in international trade regimes), or the individual values that are constitutive of global and local production and consumption systems. Where adaptation is not undertaken in response to a perceived risk (a hazard event for which a social actor is both exposed and susceptible) vulnerability will remain unchallenged. No one vision of pathway is more appropriate – all depend on the context and viewpoint of the observer.

These three visions for adaptation operate across all scales: from an individual's interior, emotional world to external expressions and actions that stretch from the household to international regimes. The result is a perplexing mixture of intention and action moving between scales and over space and time. Some analytical purchase on these complicated dynamics may be found by considering scales as nested. Nested systems allow for change at one scale to influence (or restrict) that at others. Thus, for example, transformative change in a political regime could open up new space and opportunities for local transitions and resilience in a dynamic physical environment. Building resilience can provoke reflection and be upscaled across a management regime enabling transitional and potentially transformative change, but it could also slow down more profound change as incremental adjustments offset immediate risks while the system itself moves ever closer to a critical threshold for collapse. On the ground this results in mosaics of adaptation generated from the outcomes of overlapping efforts to build (and resist) resilience, transition, local transformative change and remaining unmet vulnerabilities. Mosaics can change over time as underlying hazards and vulnerabilities as well as adaptive capacity and action are driven by local and top-down pressures.

The following sections examine each of these three modes of responding to climate change.

Table 4.1 Modes of responding to climate change

	Resilience	*Transition*	*Transformation*
Goal	Functional persistence in a changing environment.	Realise full systems potential through the exercising of rights within established regime structures.	Reconfigure the structures of development.
Scope	Change in technology, management practice and organisation.	Change in practices of governance as rights are exercised.	Reform in overarching political economy, cultural norms or scientific paradigm.
Policy focus	Resilient building practice; use of new seed varieties to make businesses/livelihoods resilient.	Implementation of legal responsibilities by private and public sector actors and exercise rights by citizens.	New political discourses.
Dominant analytical perspective	Socio-ecological systems, ecology, engineering.	Governance and regime analysis.	Discourse, ethics and political economy.

Resilience

> The ability of a social or ecological system to absorb disturbances while retaining the same basic structure and ways of functioning, the capacity for self-organization, and the capacity to adapt to stress and change.
>
> (IPCC, 2008: 880)

The IPCC definition of resilience, presented above, is forward-looking, placing emphasis on capacities rather than outcomes of self-organisation and social learning. Within this, adaptation is positioned as a subset of resilience (along with functional persistence and self-organisation). Following on from this definition, resilience is interpreted as describing those actions that seek to secure the continuation of desired systems functions into the future in the face of changing context, through enabling alteration in institutions and organisational form. Olsson *et al.* (2006) and Nelson *et al.* (2007) have argued for the need to recognise adaptation as including more fundamental shifts including the areas discussed below as transition and transformation as subsets of resilience. Where transitional or transformatory acts at local scales enable resilience at higher scales in the nested social hierarchy this is a reasonable logic. But the framework presented in this chapter finds the distinctions between resilience, transition and transformation so central to the nature of adaptation that separate identities are proposed for these three modes of response. This conviction comes from empirical work that has revealed the contradictory nature of systems functions that may be viewed as enabling resilience or stifling transition and transformation to enhanced human security dependent on the viewpoint and positionality of the observer (Pelling and Manuel-Navarrete, 2011).

The IPCC definition points to socio-ecological systems (SES) theory in its understanding of resilience, and indeed SES theory continues to be very influential in this field. The three cornerstones of the SES construction of resilience are included in the IPCC definition as it stands: functional persistence, self-organisation and adaptation (if seen as an outcome of social learning) (Folke, 2006). The distinguishing vision of adaptation as resilience is to support the continuation of desired systems functions into the future through enabling changes in social organisation and the application of technology. Such changes are facilitated through social learning and self-organisation to enable technological evolution, new information exchange or decision-making procedures. More than this, and within the limits of bounded systems, such as development policy for a single watershed or a dairy farming business, achieving resilience may require change in values and institutions within managing organisations, and this can include challenging established priorities and power relations and potentially lead to a redistribution of goods and bads (Eakin and Wehbe, 2009). In this way adaptation as resilience has the potential to contribute to incremental progressive change in distributive and procedural justice within organisational structures. When individual cases that build resilience through internal value shifts are upscaled through government action, or replicated horizontally, real opportunities can open up for contributing to transitional or transformative change in society.

Adaptation as resilience can allow practices perceived from specific viewpoints as unsustainable or socially unjust to persist (Jerneck and Olsson, 2008). This is perhaps easiest to understand in social contexts where entrenched power asymmetries and exploitative economies are manipulated to maintain power, even when this undermines sustainability. Such outcomes are less likely when local or national decision-making is held to account, but resilience can still undermine long-term sustainability while appearing to meet the demands of adapting to climate change. This can happen when sustainability challenges are recognised but the transaction costs (including political costs) of change are perceived to be higher than doing nothing, with the least bad option being to adapt within available constraints until perceived thresholds of sustainability are breached, forcing change. For example, in the use of desalination plants to compensate for water demand, the proximate need is met but at a cost of high energy use and pollution of the marine environment. The dynamism of climate change and the unpredictability of local impacts provide the additional rationale of uncertainty to justify resilience as the preferred form of adaptation.

SES theory emphasises that ecological and social systems are inextricably linked and that their long-term health is dependent upon change, including periods of growth, collapse and reorganisation (Walker *et al.*, 2006). Both a strength and weakness of SES is its presentation as an apparently value-neutral, realist epistemology, a product of its origins in systems theory. This has produced a rational and structured framework for understanding human action, one that is particularly attractive to climate change research in offering an approach for integrating human and environmental elements into the quantitative modelling of future scenarios under climate change (Jannsen *et al.*, 2006). But two limitations are inherent in this theoretical framing of adaptation. First, while power is acknowledged, the SES literature is infused with a sense of technical optimism that can downplay the contested character of social life and socio–nature relations. The messiness of adaptation decision-making (O'Brien, 2009) is not easily captured. Apparent value neutrality conspires with technical optimism to emphasise technological innovation and efficiency over critical analysis that might place more weight on the political economy and cultural root causes of risk and its perception. In this way SES theory has been criticised for a weak integration of social science theory and a tendency to allow for an over-simplification of complex social phenomena (Harrison, 2003; Jannsen *et al.*, 2006). Second, and related, both approaches focus on relational social space but limit analysis to the outer world of interactions among individuals, groups and institutions. Inner worlds of emotion and affect – value, identity, desire, fear – that give shape or meaning to as well as being drivers for public actions, including adaptation choices (Grothmann and Patt, 2005), are difficult to include.

Transition

Transitional adaptation is expressed through incremental change to social (including economic, political and cultural) relations. Transitional acts can describe both those that do not intend or do not result in regime change, but do seek to implement innovations and exercise existing rights within the prevailing order. Transitional adaptation is therefore an intermediary form of adaptation. It can indicate an extension of resilient adaptation to include a greater focus on governance, or an incomplete form of transformational adaptation that falls short of aiming for or triggering cultural or political regime change. From an empirical perspective intent is as important as outcome in indicating transitional, resilient or transformational adaptation. Not all transitional actions will achieve the intended outcomes but they nonetheless reveal critical capacity through intention.

The association of innovation with rights makes the social dimension of adaptation explicit, even when there is no observed change to regime form. This is especially true in those social contexts where rights may have been dormant or suppressed under the prevailing social system. Where this is the case, their invocation can generate transitional social change. There are many examples of this from the environmental justice movement where the asserting of existing legal rights is seen as a method for reducing vulnerability to industrial pollution, and at the same time reinforces these same rights (as well as building other capacities in local social organisation, confidence, etc.) (Melosi, 2000; Agyeman *et al.*, 2003).

The multiple scales at which social systems operate means that transition can theoretically be implemented and observed operating at the levels of individual, local, community or regime systems, and in more complicated analysis across these scales. Where adaptations and associated rights claims involve multiple levels of actor (e.g. a squatter community and municipal government) transitional adaptation demonstrates the importance of multi- or cross-scale analysis.

Both transition and transformation indicate adaptation in cultural and/or governance systems. For transitional adaptation reform is incremental, undertaken at the level of individual policy sectors or specific geographical areas. There is the potential for bottom-up, aggregate transformational change through, for example, the promotion of stakeholder participation in decision-making leading to the inclusion of new perspectives and values in emerging policy. By contrast adaptation as transformation is composed of adaptive acts that consciously target reform in or replacement of the dominant political-cultural regime as primary or secondary goals.

Transformation

Understanding the transformational possibilities of responding to climate change is helped by the notions of risk society, the social contract and human security. These are by no means the only theoretical lenses that could be brought to help examine transformational adaptation; they have been selected because together they provide a continuum for transformational adaptation that stretches from conceptualisations of development under modernity to the application of policy for national and human security. In this way they provide a landscape of ideas to help position and understand adaptation (and mitigation) that seeks to address root causes and leverage transformation in underlying social systems from individual psychology up. Like resilience and transition, transformation may be seen as an intention or as an outcome. It also operates at all scales, from the local to the international, often simultaneously and in ways that are difficult to perceive. For example, in identifying the assumptions that underlie modernity as a potential focus for transformative adaptation this directs attention to questions of identity as well as broader structural drivers. For example: through the production and reproduction of dominant cultural perspectives that emphasise and justify individualism and undermine social solidarity and collective action: a frequently identified key component of local adaptive capacity (Smith *et al.*, 2003).

The notion of a social contract can help in the analysis of crises of legitimacy that precede political regime change, and may potentially be used to avoid or capitalise upon such crises. Disasters associated with climate change triggers are one such potential trigger for political legitimacy crises – though examples of successful transformational change are rare (Pelling and Dill, 2006) and are most often associated with large-scale events impacting upon capital or other major cities where government response has demonstrably exacerbated suffering. In such cases loss of legitimacy arises when observed risks or losses exceed those that are socially acceptable. Beneath this level the impacts of disasters associated with climate change are accepted as a play-off against other gains. Of course, not everyone in society will agree (maybe not even the majority); levels of tolerance to risk or loss vary and change over time as cultural contexts evolve. In this way the social contract is kept in a tension by risk and loss (as well as opportunity) associated with climate change, and also by whose values are included in the social contract. In addition to the established social divisions along lines of class, gender, cultural identity, productive sector, geographical association, etc., climate change also requires the recognition of future generations and distant interests in local decision-framing (O'Brien, 2009). The inclusion or exclusion of these voices reflects (and reinforces) the extent to which climate change is perceived to contribute to individual disasters or crises, or as a systemic threat to the human and non-human world with individual acts and risk consequences separated in time and space. This in turn shapes priorities for social responses to climate change risk and loss and leaves open the possibility of radically different responses: on the one hand those that are inward looking and defensive and implemented through national or local risk management; on the other, a more internationalist agenda that seeks to recognise the complicated responsibility of risk generation. That the interests of future generations or citizens of second countries should be allowed into this conversation fundamentally challenges established social organisation based on the nation-state.

Can responding to climate change incorporate this dynamic and be a mechanism for progressive and transformational change that shifts the balance of political or cultural power in society? Evidence for the potential of transformational change within national boundaries can be found in the slow and limited acceptance of international aid by the government of Myanmar following Hurricane Nargis. In large part this behaviour was a result of fear of the destabilising influences of international humanitarian and development actors on the regime. A policy that analysts have also attributed to the need for the ruling military elite to demonstrate its control over society – especially at a time when the impacts of the hurricane meant that its popular and international legitimacy was at crisis point; and the potential for usurping rich agricultural land from Karen ethnic minority farmers in the Irrawaddy delta where the hurricane caused landfall (Klein, 2008). Distrust by the Myanmar regime of international and especially Western and civil society actors has been a by-product of catalysing organisational reform at the regional level. The leadership of the Association of Southeast Asian Nations (ASEAN), a regional economic grouping, in responding to Hurricane Nargis has resulted in tighter regional cooperation for disaster response.

Where transitional adaptation is concerned with those actions that seek to exercise or claim rights existing within a regime, but that may not be routinely honoured (for example, the active participation of local actors in decision-making), transformational adaptation describes those actions that can result in the overturning of established rights systems and the imposition of new regimes. Efforts undertaken to contain or prevent scope for transformational adaptation are as important as the adaptive pathways themselves in understanding the relationships between climate change associated impacts and social change. For example, it is very common for the social instability that follows disaster events to be contained by state actors. This is achieved through the suppression of emergent social organisation and associated values halting the growth of altern-ative narratives or practices that might challenge the status quo, and lead to transformation as part of post-event adaptation (Pelling and Dill, 2006).

The socio-ecological systems literature has less to say about transformation than resilience and transitional change. Nelson *et al.* (2007: 397) describe trans-formation as 'a fundamental alteration of the nature of a system once the current ecological, social, or economic conditions become untenable or are undesirable'. But for many people, especially the poor majority population of many countries at the front line of climate change impacts, everyday life is already undesirable and frequently often chronically untenable. Here we come to a central challenge for systems analysis which places the system itself as the object of analysis. Resistance in a social system can allow it to persist (be resilient) despite (and in some cases because of) manifestly untenable ecological, social or economic fea-tures for subsystem components. Examining the distributional effects of adapta-tion across scales and time in this way is likely to become a key research agenda as adaptation becomes more commonplace. Theoretical work on nested systems allows some purchase on this (Adger *et al.*, 2009), but is very difficult to develop empirically. The point at which such tensions lead to challenges for the over-arching regime serve as tipping points for transformation. Tipping points that Nelson *et al.* (2007) point out can be driven by failures that are absolute (unten-able) or relative (undesirable), so that cultural values play as much a role as ther-modynamic, ecological or economic constraints on pushing a system towards transformation.

There is scope for transformation to arise from the incremental change brought about by transitions. Subsequent claims on the existing system result in modifications at the subsystem level. Over time and in aggregate this forces an evolutionary transformation in the overarching system under analysis. It is this pathway to transformation which existing climate change literature has focused upon. With an interest in practical ways in which productive systems might transform under climate change, Nelson *et al.* (2007) describe this process as systems adjustment and include the implementation of new management decisions or the redesigning of the built environment as examples.

Where should one look to reveal the challenges and potential directions for transformative elements of adaptation and wider acts of responding to climate change? Most practical work on adaptation focuses on addressing proximate causes (infrastructure planning, livelihood management, etc.). Transformation however is concerned with the deeper and less easily visible root causes of vul-nerability. These lie in social, cultural, economic and political spheres, often overlapping and interacting. They are difficult to grasp, yet felt nonetheless. They may be so omnipresent that they become naturalised, assumed to be part of the way the world is. They include aspects of life that are globalised as well as those that are more locally configured. The former do not have identifiable sites of production and require individual and local as well as higher scales of action to resolve (Castells, 1997). The latter are more amenable to action within

Table 4.2 Adaptation transforming worldviews

Analytical frame/thesis	Root causes of vulnerability	Indicators of transformation
Risk society	Modernity's fragmented worldview. Dominant values and institutions are co-produced at all scales from the global to the individual.	Holistic, integrated worldviews including strong, sustainable development and socio-ecological systems framing of adaptation and development. Adaptation that draws together the value systems of individuals with social institutions.
Social contract	Loss of accountability or unilateral imposition of authority in economic and political relationships.	Local accountability of global capital and national governments, to include the marginalised and future generations and not bound by nationalistic demarcations of citizenship.
Human security	National interests dominate over human needs and rights.	Human-centred approach to safety, built on basic needs and human rights fulfilment, not on militarisation or the prioritising of security for interests in command of national-level policy.

Source: Pelling (2011).

national and local political space. Table 4.2 identifies three analytical frames that each reveal different aspects of domination and the associated production of vulnerability. Each points to specific indicators for transformation as part of adaptation.

The indicators of transformation identified in Table 4.2 require deep shifts in the ways in which people and organisations behave, prioritise values and perceive their place in the world. Together they help describe the features that a sustainable and progressive social system might be expected to exhibit. They operate at the level of epistemology: the ways people understand the world. Surface – transitional – changes are already observable; for example, in the uptake of socio-ecological systems framing in adaptation and more widely in natural resource management. However, transformation speaks to much broader processes of change that encompass individuals across societies, not only specific areas of professional practice, though such enclaves may yet prove to be the niches that lead to profound societal change. More tangibly, transformation that moves beyond intention also unfolds at the level of political regimes. Here the root causes of vulnerability are made most visible when latent vulnerability is realised by disaster. The post-disaster period is an important one for understanding the interplay of dominant and alternative discourses and organisation for development and risk management, and is examined below.

Conclusion

Stanton *et al.* (2008) argue that climate change calls for a new macro-economic vision: a vision that can move beyond energy-intensive growth. The existing worldview is stuck on a 'default setting' seeking to grow our way out of climate change and its attendant risks. The inequalities of extraction, and concentration of wealth that results from this pathway for development perversely increase collective vulnerability, while simultaneously accelerating climate change associated (and other environments) hazards.

Escaping from this circularity trap requires a recognition that responding to climate change needs not only to consider adapting to produced risks but also questioning how such risks are produced by dominant development pathways. Yes, mitigating greenhouse gas emissions is key, but also important are efforts seeking to resolve the detrimental consequences of land use change, constraints on education and health, and the undermining individual self-esteem that are everyday by-products of ongoing industrial and post-industrial development. Faced with this challenge, how might research and policy development on adaptation move forward? Four priorities for research are proposed that can help to better frame adaptation as a development problem.

Diversify the subject and object of adaptation research and policy

Early work on adaptation has rightly focused on a tightly bounded object for research and in so doing has succeeded in contributing to a clearly defined domain for policy. But if we see adaptation as a social as well as a technological phenomenon then there is a need to extend from this core. The object of analysis necessarily broadens from the behaviour of individuals and their constraining institutions to include organisations, governance systems, and national and international polities. In parallel the subject of analysis extends from economy and technology to include cultural, social and political opportunities, play-offs and costs of adaptive options. Importantly it is in the interaction of different worldviews and priorities established from viewing adaptation through these contrasting lenses that the richness of adaptation policy, potential conflict and scope for coordinated and progressive, sustainable development could emerge.

Focus on social thresholds for progressive adaptation

Thresholds mark the tipping points from one systems state to another, and have been recognised in climate science and also through the concatenated impacts of climate change. Less work has been undertaken on thresholds between different stages of adaptation. Research on coping has long recognised the staggered nature of household responses to risk as economic pressures cause, first, non-productive and, second, then productive assets to be expended and finally see the dissolution of households and migration as hazard impacts and vulnerability increase (Pelling, 2010). The parsimony rule in cybernetics (Slobodkin and Rappaport, 1974) presents similar guidance: that action requiring the least expenditure of resources will be undertaken first. But both coping and cybernetics focus on ex-post adaptation; less is known about stages in proactive adaptation.

As climate change impacts are felt through ever increasing multiple, direct and indirect pathways, often without being recognised, critical thresholds will be those that set the broad scope of what is possible through adaptation, and here the distinctions between resilience, transition and transformation are potentially helpful.

Recognise multiple adaptations: the vision effect

The interaction of multiple simultaneous adaptations has been recognised across the scale when, for example, household adaptations are undermined or enhanced by local government action. But this is only one axis around which adaptation and efforts to shape adaptive capacity can interact. The competing values that underpin adaptation such as resilience, transition and transformation indicate a 'vision effect' operating alongside the scale effect. This points to horizontal as well as vertical competition and complementarities in adaptation. This axis in large part explains the observed divergence between policy intention (policies) and emergence (self-organised activity) identified (Sotarauta and Srinivas, 2006) during the implementation of policy to support or enact adaptation: a gap that reveals tensions between the actions and values of competing adaptive strategies. The vision effect also helps explain difficulties in replicating, scalingup and mainstreaming innovations that may be set within wider, contradictory visions of adaptation – local efforts at transformation will have most difficulty being mainstreamed if higher levels of governance construct adaptation as an act of resilience.

Link internal and external drivers of adaptation

Shifting thinking on climate change from an external process to one unfolding as part of the co-evolution of humanity and the environment makes it more important to understand internal – cognitive and cultural – drivers for adaptation. These are no longer fringe interests but part of the nexus of internal and external drivers that shape the 'who, where and when' of adaptive capacity and action. The possibility that different adaptive initiatives could be in competition and lead to risk shifting between social groups and to non-human lives or future generations makes it all the more important to understand the deep psychological and cultural pressures that shape the propensity for different social groups to undertake particular adaptive strategies (including those that to the outside observer may appear to be self-limiting or detrimental to individual well-being).

The aim of this chapter is not to assert that adapting to climate change can itself resolve long-standing social and socio-ecological challenges brought about by dominant modes of development. Indeed, adaptation is at core a technical pursuit aimed at reducing the impacts of climate change, and the majority of activities will likely focus on local and proximate causes. These are more visible, amenable to local action and less politically charged – making them more acceptable to local actors at risk as well as those benefiting most from the current social contract. But this chapter does argue that this is not a sufficient analytical lens to understand the relative positions of adaptation, risk and development – nor for what adaptation could stimulate. Configuring climate change as a problem of as well as for development has significant knock-on effects for the logic of response. Discourses of development will need to adjust as much as local flood management regimes. With this comes the responsibility for academia to study the ways in which interest groups present and attempt to legitimise development as discourse and its local manifestations, there is potential for new political subjectivities to be created and for political space to open, and so new constellations of power may emerge (or be blocked). Where, when and how emergence of this kind takes place, and to what effect, are likely to be key questions for the coming decades as climate change associated costs and adjustments rise in political and popular attention.

References

Adger, W.N., Lorenzoni, I. and O'Brien, K.L. (eds) (2009) *Adapting to Climate Change: Thresholds, Values, Governance*, Cambridge: Cambridge University Press.

Agyeman, J., Bullard, R. and Evans, B. (2003) *Just Sustainabilities: Development in an Unequal World*, London: Earthscan.

Burton, I., Challenger, G., Huq, S., Klein, R. and Yohe, G. (2007) *Adaptation to climate change in the context of sustainable development and equity.* IPCC Working Group II contribution to the Fourth Assessment Report, Cambridge: Cambridge University Press.

Castells, M. (1997) *The Information Age: Economy, Society and Culture*, Oxford: Blackwell.

Eakin, H.C. and Wehbe, M.B. (2009) Linking local vulnerability to system sustainability in a resilience framework: two cases from Latin America. *Climatic Change* 93: 355–377.

Folke, C. (2006) Resilience: the emergence of a perspective for social–ecological systems analyses. *Global Environmental Change*, 16 (3), 253–267.

Grothmann, T. and Patt, A. (2005) Adaptive capacity and human cognition: The process of individual adaptation to climate change, *Global Environmental Change* 15 (3): 199–213.

Harrison, N. (2003) Good governance: complexity, institutions, and resilience. Accessed from: http://sedac.ciesin.columbia.edu/openmtg/docs/Harrison.pdf.

IPCC (2008) Glossary of terms for Working Group II. Accessed from www.ipcc.ch/pdf/glossary/ar4-wg2.pdf.

Jannsen, M.A., Schoon, M.L., Ke, W.M. and Borner, K. (2006) Scholarly networks on resilience, vulnerability and adaptation within the human dimensions of global environmental change. *Global Environmental Change* 16: 240–252.

Jerneck, A. and Olsson, L. (2008) Adaptation and the poor: development, resilience and transition. *Climate Policy* 8 (2): 170–182.

Klein, N. (2008) In the wake of catastrophe comes the wiff of unrest. *Guardian*, 16 May, p. 35.

Krankina, O.N., Dixon, R.K., Kirilenko, A.P. and Kobak, K.I. (1997) Global climate change adaptation: examples from Russian Boreal Forests. *Climatic Change* 36: 197–215.

Kuhlicke, C. and Kruse, S. (2009) Ignorance and resilience in local adaptation to climate change – inconsistencies between theory-driven recommendations and empirical findings in the case of the 2002 Elbe flood. *Gaia-ecological Perspectives for Science and Society* 18 (3): 247–254.

López-Marrero, T. and Yarnal, B. (2010) Putting adaptive capacity into the context of people's lives: a case study of two flood-prone communities in Puerto Rico. *Natural Hazards* 52 (2): 277–297.

Melosi, V.M. (2000) Environmental justice, political agenda setting and the myths of history. *Journal of Political History* 12 (1): 43–71.

Moench, M. (2007) Adapting to climate change and the risks associated with other natural hazards. Methods for moving from concepts to action. In E.L. Schipper and I. Butron (eds) (2009) *The Earthscan Reader on Adaptation to Climate Change* London: Earthscan.

Nelson, D.R., Adger, W.N. and Brown, K. (2007) Adaptation to environmental change: contributions of a resilience framework. *Annual Review of Environment and Resources* 32: 395–419.

O'Brien, K. (2009) Do values subjectively define the limits to climate change adaptation? In W.N. Adger, I. Lorenzoni and K.L. O'Brien (eds) *Adapting to Climate Change: Thresholds, Values, Governance*, Cambridge: Cambridge University Press.

Olsson, P., Gunderson, L.H., Carpenter, S.R., Ryan, P., Lebel, L., Folke, C. and Holling, C.S. (2006) Shooting the rapids: navigating transitions to adaptive governance of social-ecological systems. *Ecology and Society* 11(1). Accessed from www.ecology-andsociety.org/vol. 11/iss1/art18/.

Pelling, M. (2010). The vulnerability of cities to disasters and climate change: a conceptual introduction. In H. Günter Brauch, Ú. Oswald Spring, C. Mesjasz, J. Grin, P. Kameri-Mbote, B. Chourou, P. Dunay and J. Birkmann (eds) *Coping with Global Environmental Change, Disasters and Security – Threats, Challenges, Vulnerabilities and Risks*. Hexagon Series on Human and Environmental Security and Peace, Vol. 5, Berlin; Heidelberg; New York: Springer-Verlag, pp. 547–558.

Pelling, M. (2011) *Adaptation to Climate Change: from Resilience to Transformation*, London: Routledge.

Pelling, M. and Dill, K. (2006) Natural disasters as catalysts of political action. ISP/NSC briefing paper 06/01, Chatham House, London.

Pelling, M. and Manuel-Naverrete, D. (2011) From resilience to transformation: exploring the adaptive cycle in two Mexican urban centres. *Ecology and Society* 16 (2): 11. Accessed from www.ecologyandsociety.org/vol. 16/iss2/art11/.

Pelling, M. and Wisner, B. (eds) (2008) *Disaster Risk Reduction: Cases from Urban Africa*, London: Earthscan.

Slobodkin, L. and Rappaport, A. (1974) An optimal strategy for evolution. *Quarterly Review of Biology* 49: 181–200.

Smit, B. and Wandel, J. (2006) Adaptation, adaptive capacity and vulnerability. *Global Environmental Change* 16 (3): 282–292.

Smit, B., Burton, I., Klein, R.J.T. and Wandel, J. (2000) An anatomy of adaptation to climate change and variability. *Climatic Change* 45: 223–251.

Smith, J.B., Klein, R.J.T. and Huq, S. (eds) (2003) *Climate Change, Adaptive Capacity and Development*, London: Imperial College Press. Accessed from ebooks at http://ebooks.worldscinet.com/ISBN/9781860945816/9781860945816.html.

Sotarauta, M. and Srinivas, S. (2006) Co-evolutionary policy processes: understanding innovative economies and future resilience. *Futures* 38 (3): 312–336.

Stanton, E.A., Ackerman, F. and Kartha, S. (2008) Inside the integrated assessment models: four issues in climate economics. Working Paper WP-US-0801, Stockholm Environment Institute. Accessed from http://devweb2.sei.se/us/WorkingPapers/WorkingPaperUS08–01.pdf.

Walker, B., Salt, D. and Reid, W. (2006) *Resilience Thinking: Sustaining People and Ecosystems in a Changing World.* Washington, DC: Island Press.

Wisner, B., Blaikie, P., Cannon, T. and Davis, I. (2004) *At Risk: Natural Hazards, People's Vulnerability and Disaster*, London: Routledge.

Part II

Resilience and the power–knowledge interface

5 Paradigm shift in US climate policy but where is the system shift?

Marcus Carson

Introduction: a 'sea change'?

The American national elections of 2008 carried with them a fundamental shift in US climate politics. Characterized by then-UNFCCC chief Yvo DeBoer as a 'sea change' in US climate politics (Román and Carson 2009), that shift was reflected in the ways in which majorities in both Houses of Congress conceptualized climate change as an environmental, social and policy problem. It was not so much a dramatic reversal of position, however, as a logical extension of incremental developments in Congress over the preceding few years. The establishment of the House Select Committee on Energy Independence and Global Warming (in 2007), a steady increase in the number of Congressional hearings on problems related to climate change, important bipartisan support from Republican heavyweights in the Senate, and other developments all contributed to nudging climate higher up the Congressional agenda. The election brought a newly enlarged chorus of voices noting agreement on the basic problem of anthropogenic climate change, on the urgent need to act to reduce emissions, and on the need to prepare for changes already in the pipeline. The 'tip' produced by the election was a culmination of gradual developments long underway.

The biggest news, of course, was the paradigm shift embodied in the election of the new president. The views of the Obama Administration represented a monumental departure from the belief system guiding its predecessor's response to climate change, and this vastly different perspective on the causes and consequences of climate change was reflected in the Administration's public statements and its early actions. A series of executive orders set the Environmental Protection Agency back in motion, key posts were filled with people who were both highly knowledgeable about the science of climate change and deeply committed to action, and other steps were initiated towards realizing a domestic and international agenda that included climate change as a central policy priority.

The environmental optimism of early 2009 was buoyed up by confidence that this cognitive shift would be accompanied by systemic changes which would convert that thinking into reality. Most observers, particularly those from outside the USA, focused attention on anticipated Congressional legislation. With both candidates agreeing with the science and advocating strong action, climate change had not been prominent in the presidential election. Moreover, the financial crisis emerging late in the election campaign pushed virtually everything else back on the agenda. Immediately prior to the financial crisis, a comprehensive review of public opinion data concluded that:

> an overwhelming majority of Americans now believes that global warming is happening, that humans are at least partly responsible for causing it, and that the net effects will be harmful. A majority favors starting immediately to reduce greenhouse gas emissions and is willing to pay at least some higher costs for energy and even higher taxes if they directly enable emissions reductions.
>
> (Bowman 2008: 7)

Still, the opinion data also contained troubling warning signs, with quite strong partisan differences and significant regional variation (Rabe and Borick 2008; Leiserowitz *et al.* 2009), and, in the teeth of a deep financial crisis, a worrisome drop to only 30 per cent of Americans rating climate change as a top priority (Pew 2009). Nevertheless, the national, political chess board seemed set for significant action even if progress would require overcoming many long-standing obstacles (Román and Carson 2009).

By 2011, the much anticipated systemic shift had not only failed to materialize; it appeared ever more distant. Increasingly, organized actors opposed to climate-related policies challenge not only the policy remedies, but also the underlying science and the very premise that environmental constraints must be taken into account. Although the reluctance to support climate legislation was not exclusively Republican, opposition to climate legislation was key among the issues which the conservative Republican base and Tea Party activists rallied around, with opposition to cap-and-trade a litmus test issue. As just one example, popular Delaware Congressman (and former governor) Mike Castle – one of only eight House members to support the Waxman–Markey legislation – was defeated by the Tea Party candidate in the primary election for the Delaware Senate seat. Most analysts had predicted Castle as an almost certain winner – had he made it to the general election.

Both Congress and the American public are increasingly polarized along partisan lines (Leiserowitz *et al.* 2010). The substantial Republican gains in the 2010 elections expanded Congressional opposition, and while the Obama Administration appears to be holding to its long-term policy goals on climate and energy, the 2011 State of the Union Address indicated a shift of strategy and change in rhetorical focus to what the Administration sees as the more politically defensible territory of secure, renewable energy.

Clearly, converting idea change into social-systemic change has proven a daunting task. Even while acknowledgement of climate change as a serious social-ecological problem had deepened considerably among American political leadership following the 2008 elections, the ability to adopt and implement policy remedies has remained painfully limited.

Explaining stagnation and progress

One fundamental challenge of adopting and implementing policies for substantially reducing greenhouse gas (GHG) emissions is that it requires penetrating the heart of America's economic structure with changes certain to reconfigure the list of societal winners and losers. Such a reshuffling represents a tangible threat to those who see themselves as potential losers, prompting them to invest resources and mobilize to protect their privileged position (Schnaiberg 1980). As a result, essential elements of the shift towards greater environmental and energy sustainability are being frustrated by entrenched patterns of behavior and by influential societal actors working to reinforce and preserve those patterns to their own benefit. A powerful constellation of free market advocates, fossil fuel money and disgruntled citizens has thus thwarted the decisive shift away from fossil fuels that national legislation would likely have helped set in motion.

The dynamic outlined above may be read as a struggle between competing constellations of material interests – one central to a status quo fueled by GHG emitting fossil energy sources, the other guided by a more enlightened self-interest also pursuing reduced climate impacts. In theoretical terms, this grand struggle is driven on the one hand by a *Treadmill of Production* (Schnaiberg 1980; Schnaiberg *et al.* 2002) and by pressures for *Ecological Modernization* (EM) (Spaargaren and Mol 1992; Mol and Spaargaren 2002) on the other. *Treadmill* theories emphasize factors that reinforce path dependencies – the logic of capital accumulation and growth, sunken costs, and organized interests linked to existing institutional arrangements that define power relationships. These make the established order remarkably robust even in the face of serious challenges. While remaining within a market-based logic, EM theories highlight factors that generate significant departures from established non-sustainable pathways, emphasizing breakthroughs within individual firms, policy sectors, or particular political entities, and at the societal level. They are useful in identifying potential pathways to fundamental reforms as well as the logics by which they work. Breakthroughs may be catalyzed by crisis, mass mobilization, or entrepreneurial actors and economic self-interest – or more typically some mix of these. The complementary mechanisms highlighted by these theories helps explain recent developments in US climate policy (Carson and Román 2010). As they are employed in this chapter, such explanations focus on change processes; they remain agnostic on the deeper question of whether market logic is ultimately compatible with environmental sustainability.

While these contrasting theories offer important insights, they risk overlooking the importance of the discourses through which society–ecology relationships are conceptualized and structure the perception of ideal and material interests. This chapter argues that change pathways often entail stepwise, iterative processes in which ideas often lead, and subsequently inspire and contribute to corresponding systemic changes. This suggests that social struggles take place in analytically distinct, but interrelated and interactive arenas. One is cognitive and can be traced in competing political/policy discourses and the way they define, explain, and propose to remedy social problems. These discourses comprise belief systems that can be analyzed using the concept of policy paradigm. The other change process entails enacting (or resisting) social-structural changes, including changes in institutional rule regimes, organizational changes and social practices that follow on from the paradigmatic blueprint. During periods in which fundamental change is possible, both discourse and rule structure are likely to be arenas for intense contestation as societal actors mobilize to define and construct a new order – or defend the established order.

Examples of institutional rule changes linked to climate policy would include building research capacity, investments to improve infrastructure, requirements for inclusion of renewable energy (renewable portfolio standards), and pricing GHG emissions via taxes or emissions trading. The political capacity to endure the disruptions involved in implementing these kinds of policies relies upon acceptance of an alternative paradigm that explains why this is essential – for example, acknowledging that we are now pressing against the limits of our planet's carrying capacity on multiple fronts – and also helps to reconcile societal goals previously understood by many as being mutually exclusive or antagonistic. One clear example of goals conflicts is the perceived tension between economic growth/competitiveness goals and those of reducing climate impacts. This conflict is often central to market-liberal accounts of ecosystems–society relationships.

Neo-institutional theory and policy change

This analysis is anchored in neo-institutional theories that emphasize the rule-based nature of social interactions (Burns and Flam 1987; North 1995; Brinton and Nee 1998). From this perspective, social systems are comprised of complexes of rule-based structures that facilitate some actor choices and constrain others. Neo-institutional theories emphasize factors that contribute to stability and/or change – above all through processes in which organized actors mobilize to challenge or defend an established order. Formal policy-making around problems such as climate change represents one such set of processes; the entrepreneurial activities of market actors or social movements represent another. In its handling of the structure/agency dichotomy, neo-institutionalism recognizes cultural structure in concepts such as belief systems and policy paradigms, and recognizes social structure in terms of systems of governance norms and procedures, laws and regulations, markets, etc. Agency is exercised in this analysis primarily by organized actors pursuing ideal and/or material interests. Structure within the agency dimension is conceptualized in terms of organizational actors and in their constellations of relationships. The basic dimensions of this theoretical framework therefore seek to take into account cultural factors (in terms of policy paradigmatic beliefs), structural factors (as institutionalized rule systems, and as exogenous structure such as natural resource limits) and agency factors (as interacting constellations of organized actors) (Carson et al. 2009).

Policy paradigm: Often used metaphorically to denote particular modes of thinking and acting, the policy paradigm is here applied as an analytical concept for examining belief systems addressed to collective problem-solving. The paradigm was adapted initially from Thomas Kuhn's scientific paradigm to questions of politics and public policy by Dunlap and Van Liere (1978), Hall (1993) and others, and further developed by Burns and Carson (2002), Carson (2004), and Carson *et al.* (2009).

A paradigm provides policy actors with a shared conceptual model for collective problem-solving by offering a broadly coherent complex of assumptions and principles, simplifying metaphors, and interpretive and explanatory discourses. As a form of belief system, it helps sift through and organize information in a chaotic environment in which actors engaged in making or influencing public policy are frequently required to make decisions with limited expertise and incomplete or contradictory information. The paradigm helps define what does or does not constitute a problem, the kinds of actions and institutional arrangements considered preferable or to be avoided, and suggests the boundaries between right and wrong. It also defines the kinds of expertise considered trustworthy and knowledgeable, which authorities should be responsible for implementing preferred remedies, and procedural rules for arriving at legitimate decisions (Carson *et al.* 2009).

The social movements literature makes use of the related concept of framing (Snow *et al.* 1986). In this analysis, a policy problem is framed, or situated, in relation to the broader set of paradigmatic beliefs and understandings. In its relationship to particular institutional arrangements or a particular regime, the policy paradigm serves as a blueprint for institutional design – an idealized model of the kinds of phenomena that constitute public policy problems, and related policy goals and remedies.

Particular policies often compete with other policies for space and priority. Most policies serve numerous purposes and produce multiple outcomes, not all of which are intended. The policy paradigm provides a frame of reference for understanding and explaining these relationships. Here we are especially interested in policy principles that are on the one hand defined as being mutually exclusive, or on the other defined as mutually reinforcing. The obvious example here is the tension between environmental/climate protection on the one hand, and jobs and economic growth on the other. Yet, conflict between high-priority policy goals is not an inherent feature of those goals, but is rather a function of how policy problems are defined, how goals are established, what means for achieving those goals are selected and how they are implemented.

Paradigm shift, system shift

A paradigm shift may be said to have occurred when core guiding principles and policy priorities are replaced with new ones (Hall 1993), including when the rank ordering of policy priorities is changed in ways that take societal developments in a fundamentally different direction. The articulation of a new paradigm is not sufficient for a paradigm shift to be said to have occurred; concrete evidence of a conversion such as actions taken in line with the new paradigm that would not have been taken previously must be discernible (Carson *et al.* 2009). Over time, incremental steps may contribute to the full institutionalization of a new paradigm – a system shift.

Major shifts in public policy typically take place over time spans that extend over a decade or more (Sabatier and Jenkins-Smith 1993). Given the scale and scope of the transformation required to reduce GHG emissions, there is good reason to expect the shift from fossil fuels to also be measured in decades – barring serious crises that serve as a catalyst. Under normal circumstances, initial steps towards enacting and implementing major new policy reforms frequently emerge through the defining and establishing of new social problems (Spector and Kitsuse 1987), and strategically framing such problems in a context that makes them salient (Snow *et al.* 1986). This is very much the process we have witnessed over the past two decades or more. The reforms required by those problems as defined must also be institutionalized in the form of new rules and regulations, organizational restructuring and other significant modifications in order to achieve the behavioral and social-structural changes that are deemed essential. This multi-step process may be conceptualized as beginning with defining and establishing a problem in public policy discourse, developing, adopting and implementing suitable policy remedies, and ending with behavioral change (Keck and Sikkink 1998; Carson 2004). As we can observe from years of debate about climate science, efforts to change course away from business as usual are likely to be deeply contested at each of these steps.

In practice, the process is seldom as linear and stepwise as described above. Cultural and institutional change are parallel, iterative processes, each feeding back into and influencing the other. For example, establishing an investigatory commission would be considered an institutional or organizational change, yet it is frequently an important step in putting a new set of social problems on the map. Likewise, getting people to adhere to new rules entails internalization or belief change even after new laws are adopted.

Competing paradigms: dichotomies, goal conflict and climate change

Only a decade ago, climate policy measures to reduce GHGs were far more widely conceptualized as an inevitable threat to continued improvements in quality of life defined in terms of economic growth and development. The Byrd-Hagel Resolution (S.98), adopted by the US Senate in July 1997 prior to the Kyoto climate negotiations, provides a relevant and familiar example. The Senate announced it would not ratify a treaty without binding targets for developing economies seen as future competitors, or which 'would result in serious harm to the economy of the United States.' The expectation that any climate treaty would inevitably result in serious harm was prevalent in much of the committee testimony (Helms 1997). As is well known, these were the central points of contention articulated by the Bush/Cheney Administration in its shelving of Kyoto and its opposition to GHG reduction measures. There were other areas of contention as well. For example, the Bush Administration emphasized the apparent certainties of projections of economic damage that climate policies might generate, but forcefully sought to highlight remaining areas of uncertainty in assessments made by mainstream climate science.

Promoting the Promethean paradigm

The perception of climate policy as an economy-crusher might well have originated organically from concerns about the scale of measures required to curtail greenhouse emissions. However, its persistence in the public debate at all levels is clearly a function of its vigorous promotion (in the US, at least) by free market think-tanks and fossil fuel interests, among others (McCright and Dunlap 2003; Jacques *et al.* 2008). This set of arguments entails disagreement on virtually every front with the broad consensus on climate change as a serious, human-generated problem requiring concerted governmental response at a global level. In its purest form, this opposition goes well beyond the issue of climate change to deny environmental and/or resource limits, and environmental sustainability threats in general. In its disagreements, it provides a narrative of the source of social and ecological problems and their resolution that contains an at least superficially plausible explanation and internally coherent logic. Embedded in this narrative are claims regarding legitimate sources of expertise, which authorities should be responsible for taking action, and how and where decisions about taking action should be made.

There are many variations on this general complex of elements, characterized by Dryzek (2005: 51–74) as the Promethean Discourse. Hollander's *The Real Environmental Crisis* (2004) provides an ideal type for this discourse, articulating the entire range of core arguments pertaining to problem definition, causal relationships and acceptable remedies, and also seeks systematically to define particular actors as legitimate and others as untrustworthy. The Promethean Discourse weaves together a complex of arguments about the interrelationships between economy, society and environment, presenting a counter-movement narrative challenging sustainable development as a process that requires either planning or regulatory intervention by public authorities. As a paradigmatic model, the Promethean Discourse not only justifies and legitimizes the particular kind of economy–society–ecology relationships embodied in the free market treadmill; it also argues that there is no other alternative.

Hollander's presentation runs roughly as follows: environmental degradation is a necessary transitional phase in development – an unfortunate by-product of the early stages of economic growth. As societies develop economically and become wealthy, he contends, people naturally turn their attention to the damage that has been done and seek to correct it: 'Is it not persuasive that for decades the robust economic growth of the affluent societies has coincided with their continuing environmental improvement?' (Hollander, 2003: 15). Environmental problems, according to Hollander's logic, are first created by poverty, and then resolved by development, prosperity and technology. The remedy in this causal account of environmental problems is therefore to permit economic development to take its 'natural' course without interference by the state, since development is served best by unfettered markets. Hollander's argument rests of course on an unstated assumption that the only path to economic development is the known path taken by already developed countries; it also studiously ignores evidence of externalized costs connected with the use of common pool resources or pollution of the same.

Hollander seeks to deny legitimacy to two important groups whose expertise is brought to bear in climate and other environmental issues: the majority of scientists involved in climate-related research (especially those linked to the IPCC process), and members and supporters of environmental movements and organizations. Leaving aside his arguments about the state of climate science (which represent conventional denial arguments), he accuses the IPCC of being political rather than scientific, and thereby a tainted source of expertise on climate change. He defines 'good' scientists as those who express natural skepticism towards scientific orthodoxies such as anthropogenic climate change. But by this standard it is not scientific skepticism that distinguishes the credible climate scientist, but the choice to vocally and publicly express disagreement with the vast majority of other climate scientists. Trustworthy expertise is therefore a minority affair, to be found among those who not only find fault with current knowledge about climate change, but those who also express certainty as the necessary standard for justifying public action.

Similarly, while Hollander observes that virtually everyone is concerned about the environment, he labels as extremists and anti-technology pessimists those who express deep concern and demand remedial action. The criterion for distinguishing between good and bad environmentalists is whether they express doubts about the self-correcting capacity of markets and future technological developments, and then argue for governmental intervention to reduce or repair human-caused environmental damage.

The collection of claims articulated by Hollander and others subscribing to the Promethean Discourse constitutes a policy paradigm in that it contains an internally coherent set of causal explanations for the problems it identifies (poverty due to the lack of economic development), defines the kind of legitimate expert knowledge that can provide information about such problems, and defines legitimate means for resolving those problems. Arguments from climate contrarians include these elements in various combinations. Environmental problems are defined as a side effect both in their creation and in their resolution, but not as a matter for serious concern. They define as untrustworthy societal actors who support the consensus summarized by the IPCC (more political than scientific), and as technology-hostile and pessimistic those who demand societal action on the basis of established levels of risk that fall short of complete certainty. At the same time, it also articulates claims about climate policies as certain to generate economic havoc.

In official settings, these arguments have been promoted by people such as Oklahoma Senator James Inhofe, well known for his denial of the science and criticism of the societal response that tackling climate change would entail. Inhofe claims: 'much of the debate over global warming is predicated on fear, rather than science,' labeling the threat of catastrophic global warming the 'greatest hoax ever perpetrated on the American people.' Inhofe refers to proponents of climate action as environmental 'extremists and their elitist organizations' and argues that they use the issue for fundraising purposes (Inhofe 2005). Taking a slightly different tack, Jim Sensenbrenner, the ranking Republican on the House Select Committee on Energy Security and Climate Change, recently disbanded by the new Republican majority in the House, argued that even if climate change was a problem, the proposed remedy would be worse than the disease because it would harm US competitiveness vis-à-vis its major developing country competitors such as India, and especially China. The 2010 US Congressional election represents a resurgence of the Promethean paradigm among Republicans, reflected in widespread climate denial among new House Republicans and intensified attacks on the EPA.

Sustainability and ecological modernization

The Obama Administration articulates an account of the economy–society–ecology relationship consistent with the sustainability discourse. It acknowledges environmental limits and urges a precautionary approach. This is reflected in a variety of ways, including embracing climate change as an urgent problem and acceptance of the risk assessment summarized in the various IPCC reports. Given the significant economic and social disruptions entailed in the scale of energy transformation required to reduce climate risks, an aggressive climate policy necessarily prioritizes the environmental element of the sustainability triad – at least in the foreseeable future. The 'sea change' embodied in the 2008 Democratic election victory therefore entailed something more than simply adding climate policy to a list of policy goals; it meant seeking to assign priority to climate policy, then subsequently seeking ways to make those policies compatible with economic recovery. Notwithstanding the critique of environmental groups regarding particular Administration efforts to reconcile economy and environment, the Administration's overall effort is clearly to shift focus from the apparent incompatibility between the economy and environmental protection to arguing that they can be mutually supportive. Its policies as a whole appear to fit within the EM discourse, although it is unclear whether the EM discourse represents end goals or is seen as a pragmatic means to orchestrate change.

To summarize the essential elements of the EM paradigm guiding the Obama Administration, the scientific community is considered to be the authority on matters of climate change and alternative energy sources. Market actors are important participants, but on condition that governments revise the institutional rules of the game under which energy is produced and consumed. The move away from carbon-based energy sources is characterized as necessary to combat climate change, but is equally emphasized as a means for revitalizing the economy and securing reliable energy supplies. In this way, it seeks to flip on its head the Promethean argument that environmental regulation is tantamount to economic suicide.

In the grip of the worst economic crisis in decades, the Obama Administration maintained climate change in its top tier of priorities largely by redefining climate policy as a means for addressing other pressing problems. What was previously understood as a zero-sum proposition between climate policy and economic growth was reframed (with some measure of success) to understanding climate-related policies as facilitating economic growth, energy security and improved living standards. Beginning with the American Recovery and Reinvestment Act (ARRA), it began promoting climate policies, broadly defined, as an important tool for job creation and economic recovery. The cap-and-trade proposal as originally urged by the Administration included auction revenues as a means for reducing the deficit and further investing in renewable energy. A decisive transition from fossil fuels was also cast as a means to achieve other important goals, including improving energy security and regaining international credibility (Román and Carson 2009).

In the practical task of seeking to institutionalize an alternative paradigm, the ARRA dedicated approximately 18 percent of its US$787 billion to measures expected to reduce American climate change impacts, prompting references to a Green New Deal. Total allocations for clean energy 'green' jobs amounted to some US$115.8 billion and were predicted to rescue or generate some 3.5 million jobs. It also provided support to state and local governments for greater energy efficiency and reducing energy usage. Tax incentives targeted investments in various environmental technologies ranging from cars to housing (Foshay and Schneider 2009). An economic crisis that was initially feared as a death-knell for near-term climate policy was instead transformed into an important boost.

Following on the heels of ARRA, the March 2009 White House budget proposal, which constitutes part of the basis for Congressional budget deliberations, continued the pursuit of climate policy via specific funding proposals. The documents reflect the way in which the coupling of climate, energy and economy was embraced in the discourse of the new Administration:

> Lack of investment in the future is most glaring in the area of clean energy. For decades, we have talked about the security imperative we have to wean our Nation off foreign oil, which is often controlled by those whose interests are inimical to ours. And in recent years, a consensus has developed over the need to limit GHG emissions, which produce global warming and increase the risk of severe storms and weather conditions that might ruin crops, devastate cities, and destabilize whole regions. All of these facts are reason enough to invest in clean energy technologies. But there is an economic imperative to embrace these investments as well. The clean energy sector presents us with immense promise—to develop and dominate a new industry sector and to create high-paying jobs here at home. From new, highly fuel-efficient cars to renewable sources of power, there are a host of

emerging technologies that can spur the growth of new business while creating millions of new jobs. Our economic competitors know that. That's why they are racing to dominate these industries and to transform their economies.

(OMB 2009: 13)

A clear trademark of Obama's Administration is its effort to engage a wide range of interests, both ideal and material, in its efforts to promote environmental sustainability and climate mitigation. One essential feature of the EM discourse is its engagement of market forces (i.e. the profit motive) as an engine for change in a more ecologically sustainable direction (Dryzek 2005). This reframing has important consequences, since it provides a coherent conceptual model for how multiple problems can be remedied in ways not previously considered viable. In doing so, the new paradigm facilitates a reconfiguration of the constellations of organized actors with power to influence policy. Through the defection of former opponents of climate policies and the recruitment of new advocates through the coupling of other related concerns (such as energy security or the promise of economic opportunity), proponents of the EM paradigm have gained ground even though they have thus far fallen short of the success represented by national legislation.

Institutionalizing a new paradigm

Thus far, discussion has focused on aggregate-level phenomena. Yet, the political, economic, cultural and energy-structural diversity of the USA creates an unevenly distributed geography of obstacle and opportunity, with fundamental change at the aggregate level conditioned by changes that take place via multiple pathways and through diverse actions at the subnational level. There are numerous paths by which a new paradigm may become institutionalized, and in practice, systemic change is realized at the national level in a culmination of subnational changes. The overall dynamic of the US policy system is therefore highly complex, with tipping points influenced by political shifts, technological developments, and shifts among the constellations of power at local, regional and national levels (Carson and Román 2010).

It is precisely these kinds of diverse change that set the stage for the efforts of 2009 to 2010. Their development continues to unfold in spite of climate gridlock's return to Washington. Important examples are outlined below, with additional detail available elsewhere (see Rabe 2004; Román and Carson 2009; Carson and Román 2010; Carson and Hellberg 2010).

At the federal level, the Obama Administration continues to pursue actions that do not require Congressional approval, contributing to the institutionalization of the problem definition and remedies of the EM paradigm. Obama used administrative authority in January 2009 to grant California and other states authority to tighten tailpipe emissions regulations (Broder and Baker 2009). The EPA continues to proceed cautiously with regulating CO_2 as a pollutant under the Clean Air Act following a 2007 Supreme Court decision. Additional rules are being developed that would indirectly reduce GHG emissions because they would reduce the economic viability of the oldest and dirtiest coal-fired plants. If proposed rules restricting NO_X, SO_2, mercury and acid gases go into effect as scheduled, some 25 to 50GW of highly polluting coal-fired utility capacity is expected to be retired. Even with new coal plants, the expected net reduction in coal-fired output would be significant, lessening the national demand for coal and thereby weakening the coal lobby. EPA regulation of GHGs from stationary sources such as power plants will have a substantial impact on emissions – but only if Republican efforts to cripple the EPA remain unsuccessful.

Action and reaction geographically concentrated: Much of the considerable progress achieved in US climate policy during 2009 has built on developments at the state and regional level in spite of earlier Bush/Cheney Administration opposition. As of late 2009, the extent of state-level efforts to curb GHG emissions was striking. Most states (41) have established GHG registries – a precursor for taking steps to reduce emissions, while two-thirds (37) had completed or were developing climate action plans. A significant minority of states (17) have set GHG targets. In addition, more than two-thirds of the states are involved in one of three regional initiatives for capping emissions: the Western Climate Initiative (WCI), the Midwestern Greenhouse Gas Reduction Accord (MGGRA), and the Regional Greenhouse Gas Initiative (RGGI). The RGGI held its ninth auction in December 2010. With 13 individual US states ranking among the top 40 emitters in the world, the scale of these efforts is significant. Expanding regional collaboration greatly strengthens prospects for a national cap-and-trade system by both producing concrete experience and expanding the political constituency that supports it. In practice, it would also speed up the eventual implementation of a national system. In a joint White Paper released in May 2010, the three announced planned cooperation to standardize North American carbon-offset programs. Worth noting is that the WCI's emissions reporting is harmonized with EPA reporting requirements and that the MGGRA seeks to link its program to RGGI and WCI, and even to the EU ETS.

Individual states are also pressing forward. California's 2006 Global Warming Solutions Act is among the most important examples. Expressly intended to combat climate change, its preliminary design for mandatory cap-and-trade was released in October 2010 with rules to limit GHG emissions to become operative by January 2012. In spite of the conservative surge in the November 2010 elections, California's voters decisively pushed aside serious threats to derail these actions by defeating ballot Proposition 23, which would have suspended the legislation until employment rates dropped below 5.5 percent for four consecutive quarters – a rare occurrence in California. The proposition failed in spite of massive financial support by out-of-state fossil fuel interests and conservative/libertarian groups. Voters also elected former California Governor Jerry Brown – who has strong environmental credentials – over a Republican candidate who promised to delay implementation of the climate law. California's success in reducing unemployment rates with an emissions reduction law in force would provide important evidence to fundamentally shift the debate about the choice between jobs and climate policy.

Concluding reflections

Treadmill theorists make a strong case that scant evidence exists for widespread societal shifts towards more sustainable practices (York *et al.* 2003, 2010). US climate policy failures between 2009 and 2010 are arguably consistent with treadmill logic. However, societal shifts of this magnitude seldom if ever take place in a single sweep, especially in societies of the size and geographic diversity of the USA. Accounting for how change processes and related struggles play out requires encompassing forms of diversity and their uneven geographic distribution, the multiple levels and scales across which the contest is played out in markets and public policy fora, and the complex interconnections between climate change and other concerns.

This chapter describes change processes that play out in two analytically distinct spheres – cultural/cognitive and institutional/structural – across a multilevel context with an uneven distribution of cultural and structural characteristics. That unevenness produces pockets of deep resistance, such as conservative fossil fuel-producing states in which material interests and culture on balance reinforce one another. It also produces hot spots for change such as the more progressive California and West Coast, where, with no coal reserves and virtually no electricity produced with coal, ideal and material interests are aligned in ways that are far more favorable to moving aggressively towards renewables. Finally, it also produces areas of greater contradiction and incongruence, such as in the Midwest, where material interests and cultural understandings that inform politics may sit uncomfortably with one another. Just as such areas may tend to alternate between red and blue in political elections, they can also 'tip' in one direction or another on the climate question – depending on how majorities conceptualize the relationship between environment and society.

Tips may be triggered or facilitated by paradigmatic changes that redefine how interests and ideals align, and whether particular goals are defined as being in conflict with one another, or whether they can be mutually reinforcing. The redefinition of climate-friendly policies as a means for reigniting the American economy offers just this sort of shift. Moreover, the multi-dimensionality of climate change as an issue opens up opportunities for expanding the constellation of actors who support pricing carbon, and they may join for very different reasons. This is not a claim that paradigmatic ideas possess an independent power of their own, but rather that they provide an important basis for organized actors to define their interests and take sides, or even switch sides. This is indeed what we have seen in the formation of new coalitions such as US Climate Action Partnership (US-CAP), which includes both environmental organizations and large companies in the energy sector.

Discursive reconciliation of apparent conflicts can win support, but also requires positive evidence to travel very far. Hot spots for change play an important role in providing such evidence when it can be demonstrated that policy remedies do not produce the predicted harm, or that they produce tangible benefits.

US climate politics are clearly driven by interests, but these interests are themselves partly defined by the conceptual models embraced by the various actors. For some groups, paradigm change is indeed difficult. For market liberal think-tanks, embracing the problem of climate change would entail supporting actions that violate their most deeply held principles, and for some interests such principles are very convenient. For the coal industry, the comparatively weak promise held by carbon sequestration may leave ecological modernization looking more like extinction. For many others, the version of the ecological modernization paradigm now being widely promoted offers not only a way to reconcile what they consider to be mutually desirable goals, but also a reason to actively support that shift. Given that many major policy shifts that have taken place at the national level have had their way paved by a critical mass of somewhere near two-thirds to three-quarters of individual states, there is good evidence that the USA remains very near to a climate policy tipping point.

Acknowledgment

The author wishes to acknowledge financial support from the Swedish Research Council for Environment, Agricultural Sciences and Spatial Planning (Formas)

References

ARRA (2009) American Recovery and Reinvestment Act. Information available at www. recovery.gov/.

Bowman, T. (2008) *Summary Report: A Meeting to Assess Public Attitudes about Climate Change*, National Oceanic and Atmospheric Administration/George Mason University Center for Climate Change Communications.

Brinton, M.C. and Nee, V. (1998) *The New Institutionalism in Sociology*, New York: Russell Sage Foundation.

Broder, J.M. and Baker, P. (2009) Obama's order likely to tighten auto standards. *New York Times*, New York.

Burns, T.R. and Carson, M. (2002) Actors, paradigms and institutional dynamics: the theory of social rule systems applied to radical reforms, in J.R. Hollingsworth, K.H. Müller and E.J. Hollingsworth (eds) *Advancing Socio-Economics*, New York: Rowman & Littlefield, pp. 109–145.

Burns, T.R. and Flam, H. (eds) (1987) *The Shaping of Social Organization: Social Rule System Theory with Applications*, London: Sage.

Byrd-Hagel (1997) *Byrd-Hagel Resolution*, Washington, DC: US Government Printing Office, p. 5.

Carson, M. (2004) *From Common Market to Social Europe?* Stockholm: Almqvist and Wiksell International.

Carson, M. and Hellberg, J. (2010) *Washington Descends Deeper into Climate Gridlock, California and the States Creep Forward*, Stockholm: Stockholm Environment Institute. Policy Brief.

Carson, M. and Román, M. (2010) Tipping point: crossroads for US climate policy, in C Lever-Tracy (ed.) *Routledge Handbook of Climate Change and Society*, London: Routledge, pp. 379–404.

Carson, M., Burns, T.R. and Calvo, D. (eds) (2009) *Paradigms in Public Policy: Theory and Practice of Paradigm Shifts in the EU*, Berlin: Peter Lang.

Dryzak, J. (2005) *The Politics of the Earth: Environmental Discourses*, Oxford: Oxford University Press.

Dunlap, R.E. and Van Liere, K.D. (1978) The 'New Environmental Paradigm'. *Journal of Environmental Education* 9 (4): 10–19.

Foshay, E. and Schneider, K. (2009, 13 February). Congress approves clean energy provisions of stimulus. Retrieved February 15, 2009, from http://apolloalliance.org/rebuild-america/energy-efficiency-rebuild-america/data-points-energy-efficiency/clean-energy-provisions-of-stimulus-are-consistent-with-apollo-economic-recovery-act/.

Hall, P. (1993) Policy paradigms, social learning and the state: the case of economic policymaking in Britain. *Comparative Politics* 25: 275–297.

Helms, J. (1997) *Contitions Regarding U.N. Framework Convention on Climate Change*, Senate Committee on Foreign Relations Calender No. 120, 105th Congress, Washington, DC: U.S. Senate.

Hollander, J. (2003) *The Real Environmental Crisis*, Berkeley: University of California Press.

Inhofe, J. (2005/01/04) Senate Floor statement by U.S. Sen. James Inhofe (R-Oklahoma). Retrieved 13 November 2010 from http://inhofe.senate.gov/pressreleases/climateupdate.htm.

Jacques, P.J., Dunlap, R.E. and Freeman, M. (2008) The organization of denial: conservative think tanks and environmental scepticism. *Environmental Politics* 17 (3): 349–385.

Keck, M. and Sikkink, K. (2004) *Activists Beyond Borders*, Ithaca, NY: Cornell University Press.

Leiserowitz, A., Edward, M. and Roser-Renouf, C. (2009, 2010) *Global Warming's 'Six Americas'*, Bridgeport, CN; Washington, DC: Yale University School of Forestry and Environment; George Mason University Center for Climate Change Communications.

McCright, A.M. and Dunlap, R.E. (2003) Defeating Kyoto: the conservative movement's impact on U.S. climate change policy. *Social Problems* 50 (3): 348–373.

Mol, A.P.J. and Spaargaren, G. (2002) Ecological modernization and the environmental state, in A.J.P. Mol and F.H. Buttel (eds) *The Environmental State Under Pressure*, Oxford: Elsevier Science: 33–52.

North, D.C. (1995) Five Propositions about institutional change, in J. Knight and I. Sened (eds) *Explaining Social Institutions*, Ann Arbor: University of Michigan Press, pp. 15–26.

OMB (2009). *A New Era of Responsibility: Renewing America's Promise*. Washington DC: White House Office of Management and Budget.

Pew (2009) *Economy, Jobs Trump All Other Policy Priorities in 2009*, Washington DC: Pew Research Center for the People and the Press.

Porter, M.E. and Kramer, M.R. (2006) Strategy and society: the link between competitive advantage and corporate social responsibility. *Harvard Business Review* 84 (12): 78–92.

Rabe, B. (2004) *Statehouse and Greenhouse: The Emerging Politics of American Climate Policy*, Washington, DC: Brookings.

Rabe, B. and Borick, C. (2008) *The Climate of Opinion: State Views on Climate Change and Policy Options*, Issues in Governance Studies, Washington, DC: Brookings.

Román, M. and Carson, M. (2009) *Sea Change: US Climate Policy Prospects Under the Obama Administration*, Stockholm: Commission on Sustainable Development, Government of Sweden.

Sabatier, P.A. and Jenkins-Smith, H. (eds) (1993) *Policy Change and Learning: An Advocacy Coalition Approach*, Boulder, CO: Westview Press.

Schnaiberg, A. (1980) *The Environment: From Surplus to Scarcity*, New York: Oxford University Press.

Schnaiberg, A., Pellow, D.N. and Weinberg, A. (2002). The treadmill of production and the environmental state, in A.P.J. Mol and F.H. Buttel (eds) *The Environmental State Under Pressure*, Oxford: Elsevier Science, pp. 15–32.

Snow, D.A., Rochford Jr., E.B., Worden, S.K. and Benford, R.D. (1986) Frame alignment processes, micromobilization, and movement participation. *American Sociological Review* 51: 464–481.

Spaargaren, G. and Mol, A.P.J. (1992) Ecological modernization as a theory of social change. *Society and Natural Resources* 5: 33–52.

Spector, M. and Kitsuse, J.I. (1987) *Constructing Social Problems*, New York: Aldine de Gruyter.

York, R., Rosa, E.A. and Dietz, T. (2003) Footprints on the earth: environmental consequences of modernity. *American Sociological Review* 68 (2): 279–300.

York, R., Rosa, E.A. and Dietz, T. (2010) Ecological MODERNIZATION THEORY: theoretical and empirical challenges, in M. Redclift and G. Woodgate (eds.) *International Handbook of Environmental Sociology*, Abingdon: Edward Elgar, pp. 77–90.

6 Lessons from the urban poor

Collective action and the rethinking of development

Diana Mitlin

Introduction

The crisis of development outlined in this book calls out for a need to change the functioning of society, away from a materialist and individual orientation. Drawing from the experience of organized groups of the urban poor in the Global South, this chapter identifies potentially transferable lessons for organizing that can break through the social structures that prevent alternative – collaborative, ecologically sound and egalitarian – modes of development from flourishing. Empirical analysis has a gendered focus, highlighting the potential for an alternative value system that strengthens and extends collective identity and action.

I argue, in the presentation of this example, that addressing climate change requires new forms of relations both between citizens and between citizens and the state. The intention is not to suggest that the experiences expanded on below offer a simple solution that can easily be replicated but, rather, that they demonstrate what is possible. While these initiatives have yet to demonstrate their relevance at an appropriate scale to the hundreds of millions who are in need (although their scale regularly involves tens of thousands at the city scale), results to date suggest that it is important to understand these experiences and that they offer potential of interest to those concerned with social justice, poverty reduction, and inclusive and equitable development.

These methodologies have emerged in the absence of alternatives. The realities for the estimated 900 million low-income residents living in informal settlements demonstrate a serious failure of development. Market wage rates for the unskilled are very low and rarely enable the accumulation of assets and income security. Market outcomes for shelter and basic services are lacking in many respects. As shown by the scale of informal settlements, the state is unwilling and/or unable to provide secure tenure or access to even the most basic of services. Clientelist political relations dominate, with few state resources for most urban poor households. Ideological solutions have rarely secured significant and substantive change for many, even when they have achieved changes in the political complexion of the state. As a result of such neglect and abuse, leaders in informal settlements have coalesced around alternatives that offer ways to develop inclusive and pro-poor development options.

This chapter elaborates on approaches supported by Slum/Shack Dwellers International (SDI), which involves 16 core international affiliates. After describing this organizing effort and associated activities, I discuss how groups have nurtured gendered organizational practices, developing the knowledge that groups need to agree on preferred courses of action and building new relations with state authorities.

Rethinking agency for social justice: why alternative approaches have emerged and what characterizes these alternatives

The emergence of these methodologies reflects the failure of both the state and the market to provide safe and secure living environments. Markets cannot by themselves zone land for housing and develop the infrastructure with the security that arises from regulatory compliance, unless the state also plays its part. The state, in a context of resource scarcity, faces competing demands from powerful groups and may be unable to respond, especially if it falls back upon urban development models designed for and used in cities in the Global North. The scale of the problem reflects the failure of ideas, both from public intellectuals supporting social justice and professionals seeking to support the struggles of the urban poor.

What has failed?

The processes of democratic government have rarely provided substantive numbers of people with access to basic needs. All too often the outcomes lead to the reinforcement of clientelist political systems. On the positive side, in the absence of comprehensive state provision, clientelist relations can be an effective mechanism for some groups to secure access to the goods and services they need. Such a context provides opportunities for deprived individuals to manipulate the system to advance their own interests (see e.g. Benjamin (2000) for India and Auyero (2000) for Argentina). However, the inadequacies of such relations also have to be acknowledged. Delivery is partial and inadequate for ensuring that the same process of clientelist bargaining can be repeated in future elections within aligned neighbourhoods, and that pressure can be maintained on others to make the required political commitments in a context of resource scarcity. A further consequence is the reinforcement of vertical relations with associations of paternalistic authority. The experience of the urban poor is one of supplication, asking a more powerful other for favours over which there is little control, and subsequently being indebted for what has been offered in order to increase the likelihood of further 'gifts'. While Wood's (2003) representation of the negative impacts of clientelism is perhaps overdrawn, and there are some benefits for the urban poor from clientelism, the consequences can be widely recognized as the partial provision of an inadequate breadth of local public goods and services on a grossly inadequate scale.

In this context, alternative political 'approaches' (a term used here to describe ideology, strategy and goals) have been proposed that aim to address the short-comings of electoral democracy and its outcomes. Grassroots leaders may be active in political parties and in trade unions. In respect of the latter, however, many low-income citizens work in the informal sector, and hence cannot join unions. As discussed by Agarwala (2006), specifically in the context of urban India, such informal workers appear, to an increasing extent, to be focusing their political strategies towards the state in order to secure basic needs and consumption goods, rather than finding ways to pressure their employers. Elsewhere, political approaches may include local party affiliation and non-party-based political organizing. Party allegiances within informal settlements are varied. Arévalo (1997) describes how when the Huaycán self-managing urban community in Lima, Peru, established the Andrés Avelino Cáceres Association to organize the invasion of vacant land they had to manage relations with at least five political parties, including Shining Path. While this intensity is unusual, residents' associations frequently work alongside party political groupings.

In Asia, community leaders have been active within NGO-supported development approaches, one of the most notable of which has been Alinsky-inspired community organizing. Saul Alinsky was a labour organizer in the United States, active from the 1930s. He pioneered a methodology, namely community organizing, which has been influential in America since the 1960s. The approach has been replicated in numerous Asian countries, including Hong Kong, Thailand and the Philippines. Other substantive experiences include liberation theology in Latin America.

These and other experiences have contributed to the evolution of engagement methodologies described below. Members of SDI have refined their strategies over time, with contemporary emphasis on savings-based organizing. In so doing, community leaders are concerned with supporting mass-based organizing that can explore and define urban development alternatives which local groups can partner with the state, to finance and regulate. The following subsection outlines SDI's organizing model.

A summary of SDI's model of organizing

Beginning in the 1990s, exchanges between urban low-income and homeless people's organizations with similar values and principles resulted in growing bilateral links and an awareness of the value of international networking. By 1996, emerging federations and supporting NGOs recognized the need for an umbrella organization. Slum/Shack Dwellers International (SDI) was formed to promote and support international exchanges between member federations and to support emerging federations in other nations. Today, the network seeks to enhance the capacity of the affiliates to influence the policies and practices of international and national development agencies to be more supportive of urban poor interests. Since its inception, the network of federations that make up SDI has grown from the six founding members (South Africa, India, Namibia,

Cambodia, Nepal and Thailand) to 15 active affiliates in 2008 (with the addition of Kenya, Malawi, Uganda, Ghana, Zambia, Zimbabwe, Sri Lanka, Philippines and Brazil) (Mitlin 2008a). Exchanges are now taking place with groups in a further 18 countries, including Egypt, Angola, Mozambique, Indonesia and East Timor. The network interacts through a continual programme of international exchanges; it is held together by a mutual recognition of the multiple benefits of solidarity among urban poor groups. Over time, the network has secured access to donor funds it can allocate itself (rather than being directed by donors). These funds make up the Urban Poor Fund International. The progenitor of the fund has been located both within IIED and at the international Secretariat (see Mitlin and Satterthwaite 2007). The grants have helped support the growth of the network, with more affiliates, and an increased awareness of the contribution of network activities to local activities.

The SDI model establishes federations at the city and national level. In some cases, these federations are grouped around particular tenure situations (e.g. the Railway Slum Dwellers Federation in India), but often there is a single federation. SDI also places considerable emphasis on building structures where communities develop their options, supported by professionals. This is true at the local neighbourhood level and also through the interaction of professionals and community leaders at the city level. The model is one of a small support NGO with considerable relations of trust between the two parties. This NGO then mediates support from other professional agencies.

The 15 current core affiliates within the SDI network each include a federation of neighbourhood-based people's organizations or savings schemes and a technical support NGO. The federations are formed from locally managed savings schemes, organized at a neighbourhood level, whose members are mainly women. The emphasis on local savings emerges from a commitment to strengthen social relations and social capital among disadvantaged urban dwellers living in informal settlements without legal tenure. The mechanism of savings appeals particularly to women because they can see the multiple benefits that arise from coming together in small groups and collecting available finance.

Savings scheme members (primarily women) form active local organizations that can consider how best to address their own needs and those of their families. The savings groups provide members with crisis loans quickly and easily, and enable them to accumulate a fund for housing improvements and/or income-generation investments. Some funds accumulate within the locality. As groups grow in number, they establish regional and/or national funds. Particularly significant is the collective management of money. The trust this builds within each group increases the capacity of members to work together on subsequent tangible and intangible development initiatives. Finance, rather than being a means of exclusion, becomes a trigger for the formation of strong local organizations, as women combine to find ways to protect and enhance their small change. Just under US$17 million is currently in the savings accounts across all federations, with much of this locally circulated within savings schemes as loans are given to members (for consumption, emergencies and small enterprise loans) and then repaid.

Savings schemes are encouraged to federate at the city and national level, creating institutions that can share assets and resources, thereby strengthening and extending their activities and influence and sharing risk burdens. City and national federations play a significant role in negotiating with state agencies to secure policy improvements and additional resources. As Appadurai elaborates (2001: 33), federating is a strategy to achieve political influence:

> It is a simple formula: without poor women joining together, there can be no savings; without savings there can be no federating; without federating, there is no way for the poor themselves to enact change in the arrangements that disempower them.

Each of these federations works alongside a support NGO, which is staffed by professionals who assist in a range of tasks related to grant management, technical development services and documentation for a professional audience.

The individual members, working within local savings collectives, develop the confidence and skills to identify and realize development plans. As a result of organizing among some of the lowest-income women living in informal settlements, a strong emphasis on shelter-related activities has emerged. Throughout the countries in which SDI affiliates are active, women take on most domestic and child-rearing responsibilities, often completing the associated tasks alongside home-based income-generation activities. Living at risk both of eviction and from natural hazards, and without access to basic services such as running water and toilets, means that improved shelter is a priority. Groups have developed a number of strategies to improve their shelter, including investment in tenure security and physical improvements (precedent-setting investments). Through a set of specific activities related to secure tenure, the installation of services and sometimes the construction of dwellings, members of savings schemes illustrate how they can improve their neighbourhoods, demonstrate their understanding of the costs associated with this process, and learn to develop more ambitious proposals. As the processes have grown in significance, city governments and some national governments have become interested in supporting these community-driven approaches, recognizing their potential contribution to poverty reduction and urban development. Urban Poor Funds are used to provide finance once these kinds of activities grow to a significant level.

In addition to financial mechanisms, SDI affiliates use a number of further measures to consolidate an empowerment-based development process. Savings schemes are strengthened and federations consolidated through members' visits to each other, to exchange knowledge and experiences, and these take place on a daily basis within cities. There are also frequent community-to-community exchanges between cities and between groups in different countries. These exchanges help to ensure that ideas come from the poor themselves, and that solutions are not dominated by professional theories and approaches (Patel and Mitlin 2002). The strategies used by the federations are embedded in the proven practices of the urban poor and, because the savings schemes members have worked out what to do themselves, they are able to change their strategies as circumstances change. They learn what is effective through their own experience, supported by that of other communities around them. Learning, rooted at this level, consolidates confidence in the capacities of low-income groups. Moreover, the consistent horizontal exchanges build strong relationships between peers, adding to the effectiveness of local negotiations. These exchanges may be experiential (for example, seeing how a savings group has negotiated for land), related to the development of specific skills, and/or have a political purpose by bringing politicians, officials and community members together for a visit to development activities in another country.

Federations use community-managed enumerations, surveys and maps to create the information base needed for mobilization, action and negotiation (see Appadurai 2001; Burra *et al.* 2003; Weru 2004). These surveys are part of a mobilizing strategy, drawing in residents who want to participate in a locally managed identification and verification of their shacks and plot boundaries. Managing these processes strengthens existing savings groups and creates new groups. These surveys and maps also help change the attitudes and approaches of governments and international agencies, through providing a resource to the state. They shift negotiating advantage as, in many contexts, politicians and officials recognize the federations' capacity to provide a fair and accurate information base widely accepted by residents. This is information that the local authority needs, for example, for upgrading and housing development, but generally does not have.

SDI affiliates seek a development partnership with government, especially local government. Most of the homes and settlements in which federation members live are informal, and they recognize that good relationships are essential if tenure security and improved access to basic services is to be achieved. Often, the groups squat on land that belongs to a state agency, and require the government to acknowledge their right to stay and to give them tenure. In other cases they are on private land, and need state support either to negotiate tenure or find an alternative location. Local government agencies control zoning and building regulations, often placing affordable housing beyond the reach of most citizens. For shelter improvements to be affordable, such regulations need to be renegotiated. SDI groups undertake precedent investments both to demonstrate the kinds of regulatory amendments that are required for an inclusive city and to elaborate on the scale of finance and the kinds of cost-sharing arrangements that may be necessary.

The federations are aware that governments face the problem of managing the city, including dealing with squatter settlements, some of which are located on land needed for infrastructure (such as road reserves), coping with the pressure on inadequate services, and with concerns that poor-quality settlements are often judged to compromise the image of the city. The challenges governments face draw them into an engagement with federations; often, they are open to working with federations if they are persuaded that federations can help them address such challenges. Exchanges of community residents, politicians and government staff provide a platform to explore these issues. For example, during an exchange from Malawi to India in August 2006, staff from the Housing Ministry began a dialogue with federation leaders about offering resettlement to squatters in the centre of Lilongwe. The government wished to reclaim the land for its own use, but preferred to negotiate with rather than evict the residents. After being exposed to the resettlement activities of the federation in India, the officials understood how the Malawi federation might assist them in this process.

Alternative approaches

There is evidence to suggest that SDI networks have contributed to advancing the needs of low-income residents in towns and cities across the Global South. More than 150,000 households have secured formal tenure and improved their basic services (SDI 2010). More than 100,000 of these have obtained further investment capital to improve their housing. Many more have strengthened their claim on land, negotiated improvements in services and improved local self-help activities. Evidenced progress towards inclusive equitable development takes multiple forms, including details of policy changes, the scale of investments by both communities and the state, and individual studies of particular initiatives (SDI 2010). Those interested in understanding the achievements and challenges faced by these networks may be helped by the collection of papers published in *Environment and Urbanization.*

These methodologies have emerged as a powerful way of catalysing pro-poor change at the city level. Local groups of citizens can come together and increase their influence over decision-making processes at the city level. The gains they achieve relate both to their immediate needs and to core capabilities for securing continued political inclusion. The contribution of these groups to climate change is enmeshed within broader concerns for immediate development needs, such that root causes of risk are tackled head-on, while proximate causes of vulnerability and hazard are seen as less of a priority. Thus, it is not those actions framed by climate change that make these initiatives so relevant to climate change! Rather, it is the underlying approach that promotes greater collective practice and more effective ways of planning with local authorities. Lessons from this experience are important in the building of alternatives that have the power to mitigate climate change and its attendant risks. To help identify lessons, the following subsections elaborate three aspects of work: women's participation and gendered ways of working; new forms of knowledge and alternative development solutions; and methodologies for establishing new political relations between the subaltern and political elites.

Gender

Women's gendered household responsibilities in towns and cities of the Global South focus on reproduction and the care of dependants, including the elderly (see Songsore and McGranahan (1998) for an illustration from Accra). Despite this contribution, women experience multiple forms of discrimination and disadvantage in the home, the workplace and in public life. The experience of SDI is that women are particularly drawn into savings-based organizing. Hence, these networks embed this disadvantaged group at the centre of self-organized poverty-reduction strategies and actions. Women's multiple needs are addressed through ensuring that they play a greater role in local organizations and community leadership. At times, men are suspicious and demand control of the process; however, SDI's experience is that over time their concerns are addressed, and they see the benefits of the subsequent development process. A further objective in engaging women is to change the style of community leadership, away from something that reflects existing and dominant organizational norms favouring authoritarianism towards something that is concerned with building links between peers and strengthening supportive, compassionate and empathetic responses. This goes to the heart of stimulating lived, cultural alternatives that can be the bedrock of alternative development.

The mechanisms involved in supporting a women-led process are complex. The South African federations argue that they are collecting people (not money) when they save. One interpretation of this is that they are collecting friends, and that for women, the attraction associated with the process of saving might be understood in the context of building a peer support group. Savings groups are local, offering a supportive social space with relatively low transaction costs, and with some or all of the following benefits: providing reciprocal help (borrowing food, caring for children); supporting women's reproductive tasks (for example, through assistance with basic services); being morally legitimate (helping the weakest and most vulnerable members, and with a focus on inclusion), hence enabling women to defend their participation when it is criticized within the family; helping women withstand the psychological impacts of damaging domestic relations; (sometimes) offering practical support to leave abusive relations, or at least challenging the accompanying and damaging sense of guilt; providing a physical space to get away from the ongoing demands of household tasks, even for a short period; and offering a social space in which to share ambitions and then be challenged to realize them. This peer support group is important in supporting women, in part because it provides a reference group that is beyond their immediate family. The group helps to validate and reinforce new public roles for the women participants – even if at the beginning it is only a public role within a neighbourhood organization. The emphasis on helping all of those in need in the neighbourhood improves women's social status and gives them a sense of pride in their actions. As women are supported to engage with a public and non-family role, they begin to be aware of, and further develop, existing and new skills and capacities. For example, they are likely to engage with local authorities to secure tenure and/or basic goods and services. In this work, the savings schemes provide support for ongoing practices of resistance (public and private) to current difficulties. Specific methodologies help renegotiate relationships with the conscious enactment of collective practices of information gathering and precedent setting. When asked by the author about the value of joining the federation, at a meeting on 28 May 2010, one member of a group in Omaruru (Namibia) explained:

> After independence, I did not feel the independence. But now, with the federation, I feel the independence as I also have a say in things. If I want to talk, there is someone from the federation who passes by. If there is a problem, one can call on a sister from the federation.

As outlined by Robins (2008) in the context of South Africa, transformations can be extremely slow as much political culture is not amendable to rapid change. In part this is because the culture is influenced by factors external to the local organizing process, and the broader context may continue to reinforce vertical relations, an authoritarian culture, and individual and household accumulation. Women federation leaders do not simply emerge from the federation itself, but are also made by the nature of interactions with external agencies. Through their knowledge practices, SDI affiliates seek to develop forms of neighbourhood improvement that help reinforce horizontal peer relations.

Knowledge

Appadurai (2004) draws on some of these gendered social relations in an article in which he describes how the federation in India inculcates a 'capacity to aspire', encouraging federation members to have ideas and undertake actions in an ambitious manner more often associated with higher status social classes. His conclusion illustrates the ways in which, in addition to augmenting human capacity, the impacts of the organizing models also affect the ways in which members think about and act upon their lives. Appadurai's analysis of the tools and methods highlights their importance to SDI affiliates in addressing the need for new kinds of knowledge that offer pragmatic urban development strategies that can go to scale (Appadurai 2001).

SDI processes subvert the monopoly on knowledge that is often secured by professional and academic institutions. Affiliates challenge the ability of such institutions to control knowledge, both through a demonstration of its inadequacy and the grounded alternatives that organized communities produce through the tools described earlier. Community exchanges, settlement enumerations and precedents offer new ways of understanding neighbourhood problems, and provide improved methodologies to secure tenure and access to services. The exchanges assist in communication and dialogue, the enumerations build the federations' credibility, and precedents (demonstrated real-life interventions) illustrate the kinds of activities that the federations wish to scale up. Such precedents have multiple benefits: they help the federation groups understand different options; they build local capacity both to undertake projects and to pressure the state for specific actions; they illustrate to officials and politicians what is sought; and they bring the groups into the public eye and increase their legitimacy. As such methodologies evolve through local replication, they are augmented and adjusted.

The city orientation and the intensity of exchanges at the local level have been critical to building alternative community-centred knowledge and information systems. Local groups have to come together to discuss their priorities and actions. Discussions in meetings are reinforced by less formal conversations. The emphasis on local action both feeds these dialogues with information and knowledge about what has happened and why it may have happened, and at the same time uses the product of these conversations to support the emergence of new plans and activities. However, at the same time, SDI knowledge is not tested simply by what can work in a single place, but also by which ideas other groups are interested in taking up and replicating. As ideas spread from the local to national and international levels through exchanges, knowledge is refined and expanded.

Reinvigorating governance

Savings-based organizing results in citizen-led forms of co-production (Mitlin 2008b), and there has been a small but significant literature on the subject since the concept first emerged in the 1980s. The use of co-production within the SDI is similar to the original elaboration of the idea (Whitaker 1980; Parks *et al.*, 1981; Brudney and England 1983), and emphasizes that there are aspects to the delivery of public services that cannot (easily) be accomplished by public officials because they necessarily involve changes in attitudes and behaviours. Because such attitudes and behaviours are partly a consequence of social stratification, interventions through 'vertical' authorities (i.e. the state) may inadvertently reinforce just what they are explicitly trying to avoid. Hence, too great an emphasis on control may be ineffective in policing the streets, and the local constabulary may benefit from building friendly relations with local community members so that they can work together to address violence and drug dealing (for example). In their practice, such methodologies nurture alternative horizontal authorities.

The literature on co-production elaborates how state authorities can coerce people into behaving in a particular way by threatening them with penalties, but suggests that this strategy has limited effectiveness. Rather, the literature suggests that authorities should encourage the intervention of peers. The effectiveness of peer intervention relies partly on the nature of their relationship and partly on their local presence, which is almost universally more intense than an external authority can ever hope to be. For example, peers, family members and local residents can challenge the 'first steps', preventing drug taking, rioting or encouraging responsible waste management. Such peer and horizontal interventions challenge the individual to be an active agent in making the most of opportunities, rather than a passive respondent to social hierarchies that have created their low status and associated disadvantages, and in so doing change notions of active citizenship and relations with the state.

The translation of co-production into development thinking has relied heavily on the discourse of a weak state, and in some cases a relatively underemployed citizenry. This presents the South as exceptional to a much more formal and bureaucratized Northern public goods and services model. Consistent with this, it presents Southern-based co-production as temporary – awaiting the development of a more advanced state (Ostrom 1996; Joshi and Moore 2004). Ostrom (1996) argues that low wage rates among unemployed or semi-employed community members in the Global South means that co-production may be the most effective way of organizing public services. Joshi and Moore (2004) use two particular examples, one the private sector water tanker suppliers in Ghana and the other an elite-led police watchdog in Karachi, to argue that non-state institutions have a role in supplementing public institutions, due to failures of governance and logistical capacity. They suggest that those concerned with service delivery in the Global South may want to consider these approaches further, especially in environments where public authority is unusually weak (ibid., 46). However, the contribution of SDI to co-production returns to the original conceptualization of co-production, which is that peer relations between neighbours are necessarily important in contributing to the effective delivery of some public services because they trigger and support social processes that cannot be initiated by external agencies, irrespective of their organizational capacities and the particular conditions within labour markets. With this understanding, there is greater emphasis on the kinds of synergy highlighted by Evans (1996) when he argues that there is a potential for the state to contribute to the effectiveness of local organizations' activities (in multiple ways), and that local organizations can contribute to the state. The collected papers edited by Evans (1996) in a special issue of *World Development* explore some of the diversity within state society synergies.

Where possible, local SDI groups invite politicians and officials to participate in their activities. At the city level, groups have identified areas of vacant land that are potentially suitable for homes and/or relocation from flood risk areas. Savings scheme members work with all residents in an informal area to reblock and redesign the use of space, enabling state-financed improvements. At the heart of the strategy is the recognition that self-organized communities have a critical contribution to make; and that contribution has multiple components:

- Only organized communities can bridge the informal and formal, unlike the state which will recreate the formal.
- Only organized communities can address the scale of deprivation and exclusion, and empower themselves to act with agency.
- Only organized communities can deal with local relations and manage service planning, management and delivery at the neighbourhood level.

The nature of groups that arise from a co-production process (i.e. with a strong practical orientation and grounded relations with the state) appears to offer particular benefits to the poor, extending political practice both through drawing in new groups to the organizing process and persuading the state to respond positively. Co-production (whether promoted by the state or by civil society) strengthens civil society capacity; it teaches these local groups new things and new ways of acting and, in particular, it strengthens collective practice. Co-production also provides an arena within which to challenge the systems and processes of government in various dimensions, including the concepts, techniques and rationalities through which services are delivered. As civil society gains knowledge of the processes of the state (through co-production), so it more easily occupies, on its own terms, spaces that Foucault has described as governmentality (Mitlin, 2008b). The dominance of the state over its citizens depends on processes of individualization, as fragmented citizens are subject to the operation of government; co-production challenges this process through strengthening the collective, both through extending its scale and deepening its capacity through a pragmatic engagement with service delivery.

Conclusions

The efforts to support community-led development are not new, and those described above add to many previous actions, efforts and experiences. What is different about this process? The emphasis on community ideas and capacities (as opposed to those of professionals) responds to an earlier literature, including work by Illich *et al.* (1977) and Freire (2000), while the emphasis on collective organization and a capacity to negotiate with the state has been considered by Castells (1983) and Escobar (2004), among others. Recognition of the importance of localism and the contribution of credit unions and cooperatives was remarked on by Hirschman (1984). These interventions have been influenced by this legacy as well as by a host of interventions by NGOs and committed professionals in state agencies. As illustrated above, in this case they appear to have been combined in ways that enable a significant impact.

The challenges associated with climate change require society to think about and organize itself very differently. It requires us to save resources, which necessitates both greater collectivity in resource consumption and more interdependent production systems (which will in turn require greater regulation and more sensitive management of markets). The likelihood of an inability to continue to secure economic growth will make it even more imperative to support the lowest-income and most disadvantaged citizens. At the same time, climate change is likely to result in both winners and losers, and in the case of the latter there will be further needs to address. Without an ability to identify and meet needs, inequalities and associated injustice will increase. The impacts of climate change are also associated with substantive uncertainty; there will be new, unforeseen events with significant further costs. The negative outcomes associated with these risks (for example, being relocated on to safe land that is then flooded) will reduce citizen trust in systems of governance. Such realities emphasize the need to develop stronger relations between neighbours, between neighbourhoods in the city and between citizens and the state.

More pragmatically, as global institutions and national governments increasingly recognize the risks associated with climate change, there is growing attention to funding for adaptation. At the same time, the debate about the impacts of climate change on low-income urban settlements highlights both tenure insecurity and lack of basic services. This requires the ability to re-imagine the work of SDI within the framework for adaptation funding, both because the network needs continuing financial support and because the scale of these funds mean that they are significant in influencing the nature of development and the extent to which it is community or professionally led. In terms of discourse as well as substantive development programmes and donor relations, transnational grassroots networks cannot afford not to engage with these debates and evolving practices.

The view of SDI federation members with respect to mitigation is clear. Climate change is a problem for everyone and everyone needs to change their consumption practices. They do not subscribe to the view of some Southern professionals that mitigation measures are the responsibility of the North. They believe that everyone needs to change their practices and that they should do what they can. In practice, this translates into discussions about environmentally friendly energy use, tree planting and livelihoods related to recycling. The engagement of the organized urban poor in arguing in favour of mitigation is likely to be important, if a sufficient political momentum is to build behind mitigation in the South. Politicians may prefer to blame the North, but that is unlikely to be an effective position if the underlying causes of climate change are to be addressed. If the urban poor seek to mitigate their own damage to the environment, this highlights the need for higher-income households and societies to do likewise.

Hence, these networks have a pragmatic engagement with everyday practices to address climate change. But perhaps their more substantive contribution is in nurturing alternative ways of relating, which facilitate collective effort and which seek to identify and nurture the meeting of collective interests. Their contribution is also in prioritizing inclusion and equity and in challenging exclusionary and discriminatory structures and processes. Modern-day life is associated with considerable individualism, both at work and at home, and the rise in inequalities has been widely recognized. The experiences recounted here offer an alternative.

References

Agarwala, R. (2006) From work to welfare. *Critical Asian Studies* 38 (4): 419–444.

Appadurai, A. (2001) Deep democracy: urban governmentality and the horizon of politics. *Environment and Urbanization* 13 (2): 23–43.

Appadurai, A. (2004) The capacity to aspire: culture and the terms of recognition, in R. Vijayendra and W. Michael (eds) *Culture and Public Action* (pp. 61–84), World Bank/Stanford University.

Arévalo, T.P. (1997) Huaycan self-managing urban community: may hope be realised. *Environment and Urbanization* 9 (1): 59–79.

Auyero, J. (2002) *Poor People's Politics*, Durham, NC, and London: Duke University Press.

Benjamin, S. (2000) Governance, economic settings and poverty in Bangalore. *Environment and Urbanization* 12 (1): 35–56.

Brudney, J.L. and England, R.E. (1983) Towards a definition of the co-production concept. *Public Administration Review* 43 (1): 59–65.

Burra, S., Patel, S. and Kerr, T. (2003) Community-designed, built and managed toilet blocks in Indian cities. *Environment and Urbanization* 15 (2): 11–32.

Castells, M. (1983) *The City and the Grassroots: A Cross-cultural Theory of Urban Social Movements*, London: Edward Arnold.

Escobar, A. (2004) Beyond the Third World: imperial globality, global coloniality and anti-globalization social movements. *Third World Quarterly* 25 (1): 207–230.

Evans, P. (1996) Introduction: Development strategies across the public–private divide. *World Development* 24 (6): 1033–1037.

Freire, P. (2000) *Pedagogy of the Oppressed*, New York: Continuum.

Hirschman, A.O. (1984) *Getting Ahead Collectively*, New York: Pergamon Press.

Illich, I., Zola, K., McKnight, J., Caplan, J. and Shaiken, H. (1977) *Disabling Professions*, London: Marion Boyars.

Joshi, A. and Moore, M. (2004) Institutionalized co-production: unorthodox public service delivery in challenging environments. *Journal of Development Studies* 40 (4): 31–49.

Mitlin, D. (2008a) Urban Poor Funds: development by the people, for the people. *IIED Poverty Reduction in Urban Areas, Working Paper 18*, London: International Institute for Environment and Development.

Mitlin, D. (2008b) With and beyond the state; co-production as a route to political influence, power and transformation for grassroots organizations. *Environment and Urbanization* 20 (2): 339–360.

Mitlin, D. and Satterthwaite, D. (2007) Strategies for grassroots control of international aid. *Environment and Urbanization* 19 (2): 483–499.

Ostrom, E. (1996) Crossing the great divide: co-production, synergy and development. *World Development* 24 (6): 1073–1087.

Parks, R.B., Baker, P.C., Kiser, L., Oakerson, R., Ostrom, E., Ostrom, V., Percy, S.L., Vandivot, M.B., Whitaker, G. and Wilson, R. (1981) Consumers as co-producers of public services: some economic and institutional considerations. *Policy Studies Journal* 9 (7): 1001–1011.

Patel, S. and Mitlin, D. (2002) Sharing experiences and changing lives. *Community Development* 37 (2): 125–137.

Robins, S. (2008) *From Revolution to Rights in South Africa*, Pietermaritzburg: UKZN Press.

SDI (2010) *A Decade of Innovating People-centred Development*, Cape Town: SDI.

Songsore, J. and McGranahan, G. (1998) The political economy of household environmental management: gender, environment and epidemiology in the Greater Accra Metropolitan Area. *World Development* 26 (3): 395–412.

Weru, J. (2004) Community federations and city upgrading: the work of Pamoja Trust and Muungano in Kenya. *Environment and Urbanization* 16 (1): 47–62.

Whitaker, G.P. (1980) Co-production: citizen participation in service delivery. *Public Administration Review* 40 (3): 240–246.

Wood, G. (2003) Staying secure, staying poor: the 'Faustian' bargain. *World Development* 31 (3): 455–471.

7 A suitable climate for political action?

A sympathetic review of the politics of transition

Peter North and Molly Scott Cato

Introduction

This chapter engages with the politics of climate change, multiple resource crisis and, increasingly, capitalist crisis developed in the UK and elsewhere by the Transition Initiatives 'movement', which seeks to develop a positive, prefigurative grassroots politics that 'looks more like a party than a protest march' (Hopkins 2008). The chapter argues that Transition Initiatives embody a progressive politics of climate change and resource crises that rejects the dualism of adaptation and mitigation; is connected to the geographies of responsibility for past and contemporary carbon emissions, and is collective, not individual. The chapter then reviews experiences of the practices of transition in Liverpool and Stroud, UK. Liverpool is a well-known post-industrial port city in northwestern England, while less well-known Stroud is a small post-industrial town in Gloucestershire, southwest England, with a long history of ecological activism and innovation. The chapter concludes by arguing that while the politics of transition is now an international phenomenon, questions remain surrounding the extent to which a local politics of transition can provide the motive power for a fundamental reorganisation of carbon-intensive economies. The strength of the transitioning approach lies more in the generation of visions, ideas and techniques for living in a utopian post-oil community than in its ability to make the necessary changes alone.

Introducing the transition network

'Transition Initiatives' work at a city, town or grassroots level to develop community-based strategies whose aim is to reduce dependency on oil and its carbon emissions by, over time, creating fulfilling low-carbon livelihoods in localised economies. Starting in Totnes, Devon (UK), the transition model has spread to much of the English-speaking Global North and to western Europe, and more recently there are some early stirrings of transitioning in Nigeria, South Africa, Brazil, Argentina, Chile and Mauritius. While some protest 'against' climate change, Transition Initiatives argue that (1) life with less energy is inevitable and it is better to plan for it than be taken by surprise when the inevitable energy crunch happens; (2) in a globalised, 'just-in-time' economy communities have lost the resilience they had even in the 1970s to be able to cope with energy and food distribution shocks, and (3) we have to act collectively and we have to act now at a community level to address these looming crises. The philosophy is that on the one hand informed individual action is worthwhile but too little is given the scale of the challenges humanity faces, but if we wait for governments to act it will be too late. Acting at a community level is the best scale to facilitate the building of local resilience. The unleashing of the 'collective genius' of the community is channelled into a process of 'energy descent', building ways of living in a localised, community-owned economy to wean us off an addiction to fossil fuels in ways that are more connected, more enriching, and which recognise the biological limits of our planet (Hopkins 2008). Given that all greenhouse gasses are emitted somewhere, it makes sense to limit them at source – locally (Agyeman and Evans 2004). Thus arguments move beyond conceptions of sustainable development which mean that 'business as usual' can continue, presumably indefinitely, or of resilience that stresses bouncing back to the pre-crisis situation, to using crisis as an opportunity to build something better.

Business as usual is not an option, anyway. Peak oil theorists argue that resource constraints mean that complex carbon-based resource-intensive forms of society do not have a future: they will inevitably unravel (Homer-Dixon 2006; Greer 2008). Humanity can prepare for life post-oil which could be more enjoyable, ecologically sustainable and inclusive than growth-based capitalism (Astyk 2008; Hopkins 2008; Murphy 2008). The alternatives are a technological fix, a descent into a 'Mad Max' post-oil world characterised by conflict over diminishing resources and environmental collapse, or creating lifeboats in wealthier Northern communities better able to handle the changes. Given inequality of adaptive capacity, it is feasible that the other possible pathways to transition could actually exacerbate the risks from climate change, especially in the context of increased competition, perhaps of a violent nature, over diminishing resources.

Technological fixes are seen as unrealistic, while collapse and lifeboats are seen as undesirable, for obvious reasons. Transitioning is thus a hopeful, perhaps even utopian project that looks to combine rebuilding the self-reliance of small towns surrounded by market gardens and allotments, where people make and mend clothes, know about and use local food and medicinal herbs, and live and work in communities that combine high levels of social capital and connectedness with the tolerance, diversity and interconnectedness of a globalised world. These are not inward-looking, xenophobic communities: 1940s self-reliance and one-planet living is combined with 1960s feminism, anti-authoritarianism and egalitarianism, sexual freedom, anti-racism and internationalism.

Transition Initiatives argue for the construction of localised, resilient communities as a response to what is now frequently called the threefold crisis: the exhaustion of the economy's principle feedstock, petroleum; the pressures caused by climate change and policies necessary to reduce its future impact; and the nexus of finance and growth problems represented by the global economic crisis that has been rumbling on since 2007. Increasingly, more developed Transition Initiatives look to increase the amount of food produced locally (Pinkerton and Hopkins 2009), generate more local power, and manufacture more products locally to meet basic needs. The community would generate more social enterprises, and finance itself with local money and other locally owned financial institutions (North 2010b). Sustainable housing would be provided by the community, using locally sourced and appropriate materials (Bird 2010). Localisation would mean that things would be produced where it makes most sense from a perspective of economic efficiency *and* social and economic justice, not just where they can be produced the most cheaply, subsidised by cheap fuel and externalised emissions. Local economic welfare would focus more on quality of life, good, wholesome food, time for family and friends, and providing low-carbon homes, very much the perspective of the degrowth and 'slow city' movements of continental Europe (Fournier 2007; Pink 2008).

For transitioners, the capitalist crisis, climate change and resource constraints are interconnected issues, although they are normally examined separately. Many technologically advanced solutions to climate change may come up against resource limits: for example, supplies of lithium for electric car batteries, uranium for nuclear power stations. An economy that had collapsed as a result of having run up against fundamental resource constraints and/or lack of credit might not be the best form of economic organisation to facilitate the generation of solutions to climate change. On the other hand, the sudden onset of peak oil can be smoothed by accessing unconventional hydrocarbons or through biofuels, but these responses frequently result in higher carbon emissions. The two problems need to be seen as intertwined: resource constraints suggest that the ecosystem is reaching limits to the quantity of inputs it delivers to our economies, while climate change is the result of the planet being unable to absorb the waste products of a complex, industrialised society. Resource constraint means we may not have the hydrocarbons we need to power our societies as they are currently organised, while climate change means we can no longer continue to emit as many greenhouse gasses as we do now. The capitalist crisis means elites are unwilling to fund and are intellectually not committed to ecological modernisation strategies of sufficient urgency and intensity to avoid looming catastrophe. This is a necessary intellectual understanding for any progressive local strategy.

Like many social innovations, transitioning has gone from a fairly uncritical welcome through to a stage where some of the problems are identified, at which stage critics begin to emerge. Is this something genuinely interesting, or a fad? The temptation can be to write off a social innovation prematurely once obvious problems emerge, but before they are overcome. Among the criticisms of transitioning that have emerged include concerns about the relatively globally privileged nature of many of the transitioning places and of the transitioners themselves; their exceptionally unconfrontational and liberal, if not naive, social change strategy (Trapese 2008); and their celebration of apocalyptic scenarios and over-positive perceptions of a post-apocalyptic world that could actually be incredibly brutal and oppressive (Neale 2008). They have a strong scepticism about what they dismiss as 'cornucopian' thinking, the capacity of human ingenuity, harnessed by capitalism, to produce technological solutions to the climate and peak oil crises (Albo 2007; Friedman 2008). Finally, there are fundamental concerns about the effectiveness of citizen action on peak oil and climate change without a wider anti-systemic challenge (Harman 2007). Are communities physically able to implement energy descent pathways in a globalised world where control of the means of production is in private hands, and energy is not produced locally but again by privately or by state-owned companies? Do plans sit on the shelf, or remain an impossible vision of a desirable but unattainable society?

Might transitioning be little more than an attempt to whistle in the dark – the latest in a long line of symbolic practices that make it look as if something is happening without a serious attempt to build sustainable, post-carbon economy and society (Blühdorn 2007)? Transition here may be thought of as a form of therapy in two senses: first, helping people come to terms with the bleak reality that they might lose their comfortable, resource-intensive lifestyle, while, second, reconciling them to their powerlessness in the face of huge change through frantic action that will not achieve its goal, displacing them into actions that do not challenge unsustainable power relations in fundamental ways. Is it, therefore, another attempt to co-opt us into discourses that 'we are all in it together', diverting us away from anti-systemic struggles into post-political struggles for an impossible vision of 'sustainability (Swyngedouw 2007)?

As sympathetic academics who are also engaged in Transition initiatives in Liverpool and Stroud respectively, we want to argue that there is more to it than that. There is another, more positive story that can be told. We argue that, as a social movement, Transition is engaged in a programme of knowledge production about how to deal with energy crises and climate change, creating a vision of what a post-oil world could look like that might be utopian, but in the positive sense of a method of thinking in creative ways that make alternative futures possible (Levitas 2010). We argue that transitioning mobilises people to act at a scale where they feel able to act, locally, and in ways that combine feelings of responsibility for anthropogenic climate change and resource depletion with a willingness, as globally privileged actors, to do something about it. We argue that Transition Initiatives embody a progressive politics of climate change that rejects the dualism of adaptation and mitigation, is connected to the geographies of responsibility for past and contemporary carbon emissions, and is collective, not individual.

Adapting to the inevitable or mitigating the worst excesses?

Transitioning has much to offer a progressive project aimed at building a sustainable, resilient low-carbon society and economy. Any progressive strategy for dealing with climate change must include both adaptation (adapting to changes that are inevitable given warming that is already in the system) and mitigation (minimising the amount of warming in the future). Adaptation can be addressed in a number of ways, from technical adaptations to higher expected summer temperatures, storms and floods through risk and disaster management to psychological adaptation to what an uncertain future might hold for residents. Is the family home likely to flood or be storm damaged? If so, is insurance available, or will it have to be abandoned eventually? What might the implications of continuing high oil high prices and scarcity be for family members' jobs?

Transition argues that it is inadequate to disconnect adaptation from mitigation: the focus on local resilience means that a low-carbon economy and society needs to combine both. Focusing on adaptation alone assumes that present-day consumption patterns are likely to be maintained, and that there is no appetite for large-scale cuts in consumption or changes in urban socio-economic systems. It also fails to exploit the synergies and energy efficiencies offered by developments such as a local community farm (in the Stroud case) which address the adaptation and mitigation agendas simultaneously by combining the production of local food (reducing food miles and therefore emissions associated with transport – i.e. mitigation) with adaptation to a world where peak oil means a rise in food prices.

A focus on adaptation alone assumes that growth-based capitalism has no real alternatives. Citizens will continue to consume, to drive private cars and fly, and any politics that does not provide for economic growth or which restricts consumption will be electorally unpopular and thus unable to generate the levels of national political support necessary for action to take place. Within the existing political structure the hard issues must be avoided, and all we can do is prepare for the inevitable – to compete for access to resources to enable individual security. A progressive politics must argue for more collectively interdependent, open and progressive conceptualisations of futures, and challenge systems of domination that prop up unsustainable practices, including conceptualisations of progress based on unsustainable levels of travel and consumption by elites serviced by armies of poorly paid casual service economy workers. In contrast to the individualising norms of dominant capitalism, transitioning, like other social movements, encourages communities to think about these issues, collectively, not individually, and, more importantly, to develop collective strategies that combine building local resilience with lowering energy use and lower greenhouse gas emissions.

Geographies of responsibility

Second, a progressive politics of climate change and resource limits must address the need to balance local resilience with international responsibilities. It would discourage places from thinking of themselves as lifeboats, meeting their own needs (food, power, water) as locally as possible with no concern for how less well-endowed places might cope. This approach denies contemporary geographical market integration, whereby currently geopolitically wealthy localities in ecologically favourable locations may well be able to increase their resilience to future shocks quite easily, whereas places with a poorer social and economic inheritance in a more vulnerable location may be left to cope alone with settlement patterns, economic structures and population levels which are themselves the legacy of past fossil fuel dependencies and/or colonial relations. Redistribution of wealth from favoured to less favoured places, for example, using 'Contraction and Convergence' or 'Cap and Share' models is necessary both within geographically uneven nations and internationally. The question is how far carbon and material transfers and costs are part of this transfer. The interdependence of secure and less secure places may also reveal itself through migration as climate change leads to demand for population movements, and better endowed places will need to absorb environmental refugees rather than secure their own future and leave the less fortunate to get by as best they can (Hodson and Marvin 2009: 18).

Radically cutting or avoiding consumption very locally with no attention to unsustainable practices at other scales and in other places seems at best misplaced, and at worst can seem as if the problem is being addressed locally while global processes that wipe out locally generated benefits are unchallenged, a kind of inverted fiddling while the planet burns. The Transition Movement's commitment to localisation should not be confused with a potentially xenophobic parochialism (North 2010a). It would be important from a progressive standpoint to maintain the benefits of fair trade and of international connection, if transport could be justified within ecological limits. It would also be necessary to recognise that some places might be better placed to produce certain goods and export them, that there are limits to what can be produced very locally (for example, windmills, photovoltaics, electric cars, tidal barrage power stations) and that consumers do appreciate a diversity of consumption choices. As long as the costs of transport are bearable within carbon budgets or ecological limits, localisation need not mean greater isolation (Lang and Hines 1993; Hines 2000; Woodin and Lucas 2004). Transition initiatives therefore argue for a politics of economic subsidiarity (Cato 2006): producing things as locally as possible, while maintaining the benefits of communication, connectivity and cosmopolitanism.

Climate and resource politics needs a 'geography of responsibility' to engage with local *and* with global geopolitics, with inequalities between places and with the right to development in a resource-constrained world (Baer *et al.* 2007). It needs to understand the relationship between places and economies, responsibilities between local places for global problems, and an understanding of the effect which action in one place has on people far away and in the future (Massey 2004). Without equality of effort and sacrifice, without *just* sustainability (Agyeman and Evans 2004), there will continue to be a perception that those in the Global South who have little historical responsibility for in-the-system warming and who are sometimes suffering from the effects of climate change are now expected to forgo development, letting those responsible for the problem off the hook. Any progressive strategy must ensure equality of contribution to maintaining the global environmental commons (Agyeman and Evans 2004: 160–161).

Transitioners recognise this. First, it is positive that globally privileged citizens in high-consuming countries are taking responsibility for their historical and current emissions, and doing what they can about it. Instead of arguing that their action is irrelevant on a global scale, or that there is no point in the wealthy countries acting while China and India's emissions grow, they are putting their money where their mouth is. Second, transitioning, a positive project of imagining local futures and taking steps to achieve them, is not the only form of political action which Transitioners take. Individual Transition Initiative members may also be members of environmental or development NGOs fighting for global justice, of political parties, or they may take part in non-violent direct action (North 2011).

Third, transitioning may be seen as a Northern manifestation of a more global social and political movement. John Holloway argues for social change which stresses not 'their power over us', but 'our power to act' (Holloway 2002). He argues that we should focus on developing our power to change things, using the tools we have locally, rejecting conceptualisations of capitalism controlling us such that we are stuck and will be blocked unless we are confrontational. He argues for a withdrawal of our energy from 'their' system, and for us to provide for our own livelihoods by growing our own food, setting up cooperatives, generating our own power and so on. Create free spaces, uncommodified spaces, Holloway argues. In this way, Argentine pickets set up community bakeries, workers recover closed factories to produce things which local people need (North and Huber 2004), Brazilian MST members occupy unused land to grow food and create livelihoods (Branford and Rocha 2002), and Mexican Zapatista develop alternative livelihood choices to low-paid work for export to the Global North in the *maquiladoras* (de Sousa Santos 2006). Julie Gibson and Kathie Graham have also argued persuasively for a more generative local post-capitalist politics, focusing more on possibilities, and obstacles as issues to be grappled with than metanarrratives of capitalist domination (Gibson and Graham 2006). Thus a Transitioner in a Northern city helping to set up a wind power cooperative, working on a community allotment, or helping to insulate houses is engaged in the same generative political project, developing 'our' power 'to'. Individual Transition Initiatives have projects such as Stroud's Learning from the South strand, where the experiences and practices of those in less 'developed' societies that are more closely embedded with their environment and live within limits are validated and used for inspiration. Thus their actions are not just localised 'militant particularisms' (Featherstone 2008), but consciously part of a wider struggle.

Collective not individual

Third, transition is a collective, not an individual response to the three crises. Transition does not argue for individuals to 'run for the hills' and set themselves up on self-reliant smallholdings on a North American model (Loomis 2005), or adopt individually frugal, self-reliant livelihood techniques (Astyk 2008).

Rather, transition argues that attention should be paid to the sorts of livelihoods generated, supporting democratic and inclusive economic forms such as co-ops and worker-controlled businesses, developing alternative plans for currently unsustainable or questionable forms of business, and developing public and community controlled sustainable forms of provisioning. Increasingly, the focus is on pro-actively developing a localised, community-owned economy full of social enterprises and community trusts which grow food and generate power locally. Homes should be made low carbon collectively, rather than leaving the process to individual initiative.

The climate and resource crises are thus seen as a way to harness new environmental technologies for the construction of a more convivial, democratic and inclusive post-capitalist economy (Murphy 2008; Neale 2008). Rather than lecturing citizens about their individual responsibility to cut consumption while providing flights at a fraction of the cost of train travel or poor levels of public transport such that people need a private car, transitioning aims to make it the case that a low-carbon lifestyle is obvious, easy to achieve and enjoyable. The progressive politics of transition is to ensure that a transition to a localised economy is planned and seen as an opportunity, and a descent into tribal balkanisation is avoided.

From politics to action: the lived practices of transitioning Liverpool and Stroud

We now turn to an analysis of the lived practices of transitioning in Liverpool and Stroud. As engaged and committed activists, we must of course put a health warning up here: this is our take on what we think has gone well, and our self-critique of, as much as anything, our own practice. Of course, a more dispassionate analyst might well come to different conclusions.

Liverpool

Transition South Liverpool was one of the first urban transition initiatives. Formed in November 2007 the transition message quickly went viral across the city and Transition South Liverpool quickly grew to a network of about 200 members, with 60 to 70 attending early meetings at which transition films were screened. An early discussion was how citizens could transition a port city like Liverpool. How could we affect what cars were made here, how the port operated? Would it be better to work at a small neighbourhood level, as many urban transition groups do? We decided that if we focused on South Liverpool, and refused to define our boundaries too tightly, it was 'local' enough for local projects to have some coherence and focused on the part of the city where most like-minded people lived (at the time it had two Green Party councillors), but broad enough to include as many people as wanted to be involved and to, as we put it, 'go where the energy is'. After six months or so of educational work, we held our first project day: March 2008. A number of projects were set up including a community allotment, a group who wanted to explore co-housing, an energy group and proposals for a food cooperative. Some discussions were held with the council to build connections. By August 2009, the feeling was that members were less interested in discussions and education, and more interested in practical action. Key activists from the early core group had other commitments, moved on, or became involved in local projects. Following the transition ethos of 'letting it go where it wants to go', the core group was dissolved, with decisions being taken at a monthly meeting open to all members.

Four years after its foundation, Transition South Liverpool can show some successes and has learned some lessons. It operates as a network of like-minded people exchanging information and connections, circulating ideas by email discussion lists, with members physically meeting monthly and through individual projects. Its website documents its work over the years, acting as a repository of ideas and helping cement a local reputation as an active and creative group. Specific projects that the group has catalysed include a community allotment, four co-ops, two social enterprises, active Energy and Health Groups, and work with Dingle Opportunities in one of the city's socially excluded communities. More strategic work is being undertaken through 'Low Carbon Liverpool', a partnership between Liverpool University, the Chamber of Commerce, Liverpool Vision (the city's economic development company) and Groundwork Merseyside which again emerged in part from discussions through Transition South Liverpool. In part, its successes have been predicated on a permissive approach: 'letting it go where it wants to go', 'where the energy is'. This has meant that the group has largely avoided fragmentation over disputes about what should or should not be done: members suggest ideas and take action, and are either followed or not depending on how much support there is or energy for the ideas. Others feel progress has been slow and commitment weak. The group does not have strong connections to the local authority or to the city's partnerships. It has not started work on energy descent pathways, although the energy group plans this for 2012.

Transition South Liverpool operates then as a network connecting like-minded people, passing around information, and catalysing projects and grassroots interventions as and when members have the energy for it. It is not particularly strategic in its thinking, and cannot be said to be providing any sort of community leadership in the transition to a low-carbon economy. It is one of many environmental and community groups in the city, many of which are working towards similar agendas. Its members are busy as part of these projects and groups, as well as members of Transition. Consequently, it is not possible to identify what can be specifically attributed to Transition South Liverpool, as opposed to the wider milieu. In a small town or community a Transition Initiative can have a focus, provide community leadership, and have impacts not possible in a larger city where more is going on and impacts cannot analytically be isolated and ascribed to any specific group or groups. Nevertheless, it is increasingly being seen as a grassroots player in the city.

Criticisms of transition as 'apolitical' really do not hold weight in Liverpool. Transition in Liverpool is working in a port city that has a robust and confrontational local political culture (Kilfoyle 2000), and where issues of sustainability can lose out to calls for growth given the city's entrenched poverty (North 2010c). It adopts a more critical tone towards local politics than some Transition Initiatives, and is not involved in consensual participatory processes with the local authority that it sees as 'greenwash'. In discussions at Transition meetings, people struggled with the ethos of a focus on what we can do, locally, with the tools at hand, but seemed happier criticising the council for what it 'should' or 'should not' do, or with making anti-capitalist statements (i.e. focusing on the power 'they' have over us, not on our power 'to'). There was less energy for much practical or strategic action, and, perhaps, less of an understanding of exactly how to localise an economy, and a lack of the skills to do it. Transition Liverpool, by and large, has signposted some members to the city's well-developed social enterprise support network, but has not grown as many cooperators as it would have liked, and it has not had much of an impact on how the city can transition to a low-carbon economy.

That work has been undertaken by a different group, Low Carbon Liverpool,[1] which works at a more strategic level, a level grassroots members were not confident about or skilled at working. On the other hand, Low Carbon Liverpool would probably not have had the impact it has had without the awareness-raising work undertaken by Transition Liverpool. It is too soon to know to what extent Low Carbon Liverpool will be able to influence strategic partners to make a difference to Liverpool's transition to a low-carbon economy. Time will tell.

Stroud

Located at the western edge of the Cotswold escarpment, Stroud was one of the first Transition Initiatives, but much of the activity which currently carries that label has a much longer history than the movement itself. Local people in semi-rural areas like Stroud District have always grown their own food, and this habit, which is now termed 'resilience', has persisted. Stroud has been a community leading social innovation, and particularly green social innovation, for some 30 years.

Traditionally a wool town, following the end of the Second World War, Stroud's textile industry went into a rapid decline and only one mill still remains in operation. This was a devastating economic blow, but it did result in the availability of cheap property: vast industrial sites and mill buildings were derelict and this attracted both artists and environmentalists to the town. While there is little scope for well-paid employment in the town, Stroud has access to circuits of capital, largely as a result of its good transport connections, and many local people who are involved in sustainability activities travel to Gloucester, Bristol or London to work. For some, it is a deliberate decision to acquire financial, social and cultural capital during a short working week, and to spend two or three days in unpaid work for the local community.

The strength of environmental activism in the town dates back to the late 1970s, when Gloucestershire County Council drew up plans for a ringroad that would have destroyed the traditional character of the town. Stroud Campaign Against the Ringroad defeated the plans at a public inquiry. In the 1980s the District Council's plans for redevelopment of the town centre included the demolition of its historic eighteenth-century buildings. Again a protest group successfully defended the town. Following the spate of issue-based protests, a group developed to take forward a fully fledged community-planning process, leading to the establishment of the Stroud Preservation Trust, which has protected other historic buildings, as well as a large range of local initiatives, including the strengthening of local networks and the acquisition and development of community assets. More recently, local campaigns have prevented the closure of the Stroud Maternity Hospital (2006) and the Uplands Post Office (2009). The green tinge of Stroud's politics also extends to its elections. It is the only town council in the country with a majority of Green Party councillors, and the Party also has one of the highest number of councillors on a district council anywhere in the country (6 out of 52), as well as a single county councillor. The synergies among local authority officers, elected green politicians and environmentally focused community activists in creating a sustainable community is a question ripe for further academic exploration.

The market town is the favoured economic unit of proponents of resilience, who argue that it was the natural social form for trade and provisioning before fossil fuels made possible the extended supply chains of the global economy. Although a recent survey discovered that more than 80 per cent of locally owned shops in the town sourced no more than 25 per cent of their stock locally, the local environment is ripe with possibilities. A range of local food initiatives, from Stroud Community Agriculture, the UK's first cooperative CSA, to the community allotments form the basis of Stroud's alternative food economy. A new initiative from Transition Stroud for 2011 is the Open Gardens weekend, where those who grow their own will share their learning with others who lack the confidence to begin, following the pattern of the very successful Open Homes weekends, when home-related energy-efficiency innovations are passed on in a similar way.

Another feature that may be identified as contributing to the success of Stroud in responding positively to the threats posed by globalisation is an intelligent use of information, and the sharing of that information in systems of networks. Rather than the passive, disconnected virtual networks theorised by Castells (1997), these are social networks of engaged citizens, where a diversity of views is welcomed. Within Stroud it is suggested that this networking grows out of the creative response of the artist, so that when you suggest your latest woodland sculpture or community currency scheme the response is always 'Great! Why not?' rather than 'What for?'

In the spirit of these live and receptive networks, Stroud Common Wealth has established Stroud Communiversity. Launched in 2008, the Communiversity is a deliberate attempt to formalise the sharing of learning and expertise that flows around Stroud, but can exclude those who are not already part of the network (Cato and Myers 2011). It also seeks to move beyond the local community by drawing in those from communities who are attempting to move towards sustainability and share what has already been learned. As a response to the financial crisis, Stroud Communiversity organised a series of events called 'Cuts or Commonwealth', where speakers and film showings gave local people the opportunity to learn and also share their anxieties, as well as building ideas for positive responses, including greater support for the local currency, the Stroud Pound (North 2010b: 173–182, Cato and Suarez 2011). In the wake of the increase in student fees and the exclusion of young people from higher education, the idea has emerged to create Stroud Communiversity as an HE provider, with environmentally and practically focused third-level qualifications: another idea that is floating free within the network, waiting for a group of people to find the energy and enthusiasm to make it happen. In Stroud, then, we can see considerable evidence for the existence of 'green niches' (Seyfang and Smith 2007) where new technologies and tools for low-carbon conviviality are being developed at grassroots levels.

Conclusion

Melucci (1989: 40) argues that the focus of social movement activity is not always on persuading governments or elites to make changes, but on developing alternative meanings about how society *should* be organised. The social movement acts as a 'knowledge producer' (Eyerman and Jamison 1991), creating messages about how things should be, aimed at the rest of society. Under these conditions, participation in creating that sign is sufficient. Policy change is not necessarily necessary, if the required changes are not within the limits of the system to make those required changes. Consequently participation in the social movement is not a means to an end (achieving future changes), but an end in itself.

Local, practical, prefigurative manifestations of the sort of society transitioners envisage can be seen in both Liverpool and Stroud. Does this mean that the message is enough? Avoiding climate chaos means that changes *must* be undertaken, literally, if humanity is to avoid an apocalypse. If climate change activism is to stop harm or calamity then the outcome of activism must matter. On the other hand, if peak theorists are right and if we have already passed climate change tipping points, all transition movements can do is communicate what *will* happen, and encourage at best preparedness, at worst stoic quietism in the face of an unavoidable tsunami. Under conditions of inevitable change, transitioning should be seen as an effective strategy for grassroots organising, learning by doing, and knowledge production and visioning a post-carbon society – developing our 'power to' – that needs to be balanced by confident and defiant challenges to 'their' power over us, and their unsustainable, ecocidal economic system.

As for empirical evidence for the transition to a low-carbon society, Transitioning in Liverpool and Stroud has certainly raised issues that are being discussed at a strategic level in Liverpool and by the town council in Stroud. However, it is too soon to see the extent to which local policy-makers, even if persuaded of the importance of emissions reduction and energy security, will be able to influence the investment decisions of hundreds of privately owned businesses operating in a growth-orientated market economy, or the consumption choices of millions of individual citizens, structured as many of them are into unsustainable consumption practices. Higher energy prices, emissions regulations and low-carbon technological innovations may, in time, mean that economies transition to low carbon in 'immanent', rather than 'intentional' ways, through individual investment decisions responding to high energy prices and regulation (North 2010a). But the limits of the ability of public sector actors to influence private business decisions in market economies through local economic development strategies has long been understood (Cockburn 1977; Eisenschitz and Gough 1993). We also know that it is more likely that individuals will be encouraged to act in sustainable ways when structured to do so by their daily practices (Spaargaren 2003): exhortation is never enough.

Note

1 See www.lowcarbonliverpool.com.

References

Agyeman, J. and Evans, B. (2004) 'Just Sustainability': the emerging discourse of environmental justice in Britain? *The Geographical Journal* 170 (2): 155–164.

Albo, G. (2007) The limits of eco-localism: scale, strategy, socialism, in Panitch, L. and Leys, C. (eds) *Socialist Register 2007: Coming to Terms with Nature*, London: The Merlin Press.

Astyk, S. (2008) *Depletion and Abundance: Life on the New Home Front*, Gabriola Island, BC: New Society Plublishers.

Baer, P., Atanasiou, T., Kartha, S. and Kemp-Benedicy, E. (2007) *The Right to Development in a Climate Constrained World*, Berlin: Heinrich Böll Foundation, Christian Aid, EcoEquity and the Stockholm Environmental Institute.

Bird, C. (2010) *Local Sustainable Homes: How to Make Them Happen in Your Community*, Totnes, Devon: Transition Books.

Blühdorn, I. (2007) Sustaining the unsustainable: symbolic politics and the politics of simulation. *Environmental Politics* 16 (2): 251–275.

Branford, S. and Rocha, J. (2002) *Cutting the Wire: the Struggle of the Landless Movement in Brazil*, London: Latin America Bureau.

Castells, M. (1997) *The Power of Identity*, Oxford: Blackwell.

Cato, M.S. (2006) *Market Schmarket: Building the Post-Capitalist Economy*, Gretton: New Clarion Press.

Cato, M.S. and Myers, J. (2011) Education as re-embedding: Stroud Communiversity, walking the land and the enduring spell of the sensuous. *Sustainability* 3: 51–68.

Cato, M.S. and Suarez, M. (2011) Stroud pound: a tool to map and measure the local economy, *International Journal of Community Currency Research*, forthcoming.

Cockburn, C. (1977) *The Local State: Managing Cities and People*, London: Pluto Press.

de Sousa Santos, B. (ed.) (2006) *Another Production is Possible: Beyond the Capitalist Canon*, London: Verso.

Eisenschitz, A. and Gough. J. (1993) *The Politics of Local Economic Policy: The Problems and Possibilities of Local Initiative*, Basingstoke: Macmillan.

Eyerman, R. and Jamison, R. (1991) *Social Movements: A Cognitive Approach*, Cambridge: Polity Press.

Featherstone, D. (2008) *Place, Space and the Making of Political Identities*, Oxford: Blackwell with RGS/IBG.

Fournier, V. (2007) The politics of degrowth: questions and possibilities for a sustainable future. *Local Economic Development in an Era of Dangerous Climate Change and Peak Oil.* University of Sheffield, www.liv.ac.uk/geography/seminars/seminartwo.htm.

Friedman, T.L. (2008) *Hot, Flat and Crowded*, London: Allen Lane.

Gibson, J. and Graham, K. (2006) *A Post Capitalist Politics*, Minneapolis: University of Minnesota Press.

Greer, J. (2008) *The Long Descent: A User's Guide to the End of the Industrial Age*, Gabriola Island, BC: New Society Publishers.

Harman, C. (2007) Climate change and class politics. *Socialist Review* (July–August).

Hines, C. (2000) *Localisation: A Global Manifesto*, London: Earthscan.

Hodson, M. and Marvin, S. (2009) Urban ecological security: a new urban paradigm? *International Journal of Urban and Regional Research* 33 (1): 193–215.

Holloway, J. (2002) *Change the World Without Taking Power: The Meaning of Revolution Today*, London: Pluto Press.

Homer-Dixon, T. (2006) *The Upside of Down: Catastrophe, Creativity, and the Renewal of Civilisation*, London: Souvenir Press.

Hopkins, R. (2008) *The Transition Handbook: From Oil Dependency to Local Resilience*, Totnes, Devn: Green Books.

Kilfoyle, P. (2000) *Left Behind: Lessons from Labour's Heartland*, London: Politicos.

Lang, T. and Hines, C. (1993) *The New Protectionism: Protecting the Future Against Free Trade*, London: Earthscan.

Levitas, R. (2010) *The Concept of Utopia*, London: Philip Allan.

Loomis, M. (2005) *Decentralism*, Montreal: Black Rose Books.

Massey, D. (2004) Geographies of responsibility. *Geografiska Annaler* 86B (1): 5–18.

Melucci, A. (1989) *Nomads of the Present*, London: Hutchinson Radius.

Murphy, P. (2008) *Plan C: Community Survival Strategies for Peak Oil and Climate Change*, Gabriola Island, BC: New Society Publishers.

Neale, J. (2008) *Stop Global Warming: Change the World*, London: Bookmarks.

North, P. (2010a) Eco-localisation as a progressive response to peak oil and climate change – a sympathetic critique. *Geoforum* 41 (4): 585–594.

North, P. (2010b) *Local Money*, Dartington: Green Books.

North, P. (2010c) Unsustainable urbanism? Cities, climate change and resource depletion: a Liverpool case study. *Geography Compass* 2 (6): 1–15.

North, P. (2011) The politics of climate activism in the UK: a social movement analysis. *Environment and Planning A.* 43 (7): 1581–1598.

North, P. and Huber, U. (2004) Alternative spaces of the 'Argentinazo'. *Antipode* 36 (5): 963–984.

Pink, S. (2008) Sense and sustainability: the case of the Slow City movement. *Local Environment* 13 (2): 95–106.

Pinkerton, T. and Hopkins, R. (2009) *Local Food*, Dartington: Green Books.

Seyfang, G. and Smith, A. (2007) Grassroots innovations for sustainable development: towards a new research and policy agenda. *Environmental Politics* 16 (4): 584–603.

Spaargaren, G. (2003) Sustainable consumption: a theoretical and environmental policy perspective. *Society and Natural Resources* 16: 687–701.

Swyngedouw, E. (2007) Impossible 'sustainability' and the post-political condition, in Gibbs, D. and Kruger, R. (eds) *Sustainable Development*, New York: Guilford Press.

Trapese (2008) *The Rocky Road to a Real Transition: The Transition Towns Movement and What it Means for Social Change*, Leeds: Trapese Popular Education Collective.

Woodin, M. and Lucas, C. (2004) *Green Alternatives to Globalisation: A Manifesto*, London: Pluto Prss.

8 Ecological modernisation and the spaces for feasible action on climate change

Andy Gouldson and Rory Sullivan

Introduction

Climate change is often seen as one of the most intractable problems that we face today. The complexity of the science, the social, temporal and geographical distances between cause and the effect, the scale and scope of the required response and a range of other features all conspire to make it such an intractable problem. When coupled with the limited time available to develop an effective response that allows us to avoid dangerous levels of climate change, the issue really does demand that modern societies do a lot of learning very quickly.

To some extent this has happened. Climate science – perhaps led by the natural sciences but now very definitely including the social sciences – has evolved comparatively quickly over recent years. As is evident from the expansions in the breadth and depth of the research cited by the Inter-Governmental Panel on Climate Change's first report in 1990 (IPCC, 1990) and the fourth report in 2007 (IPCC, 2007), our understanding of climate risks and impacts, and of the range of mitigation and adaptation options has advanced rapidly in the past two decades. Technologies have emerged relatively quickly too – indeed engineers have been very quick to point out that a large chunk of the climate problem can be tackled with existing technologies, stressing that there is no need to wait for major technological breakthroughs (cf. Pacala and Socolow, 2004). In addition, the economic landscape has changed rapidly – not least as arguments have been put forward that it will be cheaper to tackle climate change than not to (Stern, 2006), or that there are major opportunities in the emerging low-carbon economy (cf. Enkvist *et al.*, 2007). Politics and policy have changed as well – politicians are learning how to present the argument for transition to a low-carbon future in new ways – emphasising, for example, the various co-benefits of tackling climate change, notably by reducing dependence on unpredictable energy supplies or by creating opportunities for a new wave of low-carbon economic development. They are also developing new forms of policy and governance that seek to tackle different elements of the climate change problem. Put all of this together and it seems that within a few years we have moved from a situation where climate change was one part of a 'runaway world' (Giddens, 2002), towards a situation where we can at least envisage ways of reasserting some level of control over this aspect of the future.

As will be discussed below, these changes may all be seen as classic manifestations of the processes of ecological modernisation (EM). EM theories have been quite extensively discussed for nearly 20 years now, with the significant developments taking place throughout the 1990s and into the early 2000s (see Mol *et al.*, 2009). However, EM theories have a lot to offer as we seek to make sense of climate change and of the ways in which societies are attempting to cope with it. As has been discussed more extensively elsewhere (see Bailey *et al.*, 2011), a review of EM theories can help us to put contemporary developments in climate policies into some sort of critical context.

Within this chapter we explore these issues first conceptually, where we briefly examine the nature of debates on EM before proposing a simple framework for understanding the spaces for feasible action on climate change, and then more practically, where we explore the explanatory value of this framework at the macro and micro levels. We then consider the implications of this analysis for policy and governance, before concluding with a discussion on how this both informs and can be informed by an analysis of EM theories.

Ecological modernisation

Ecological modernisation (EM) theories focus on the relationship between environment and economy and on social capacities to recognise and respond to environmental problems (see Gouldson *et al.*, 2008; Mol *et al.*, 2009; Bailey *et al.*, 2011). In essence, ecological modernisation argues that a mutually antagonistic relationship between environment and economy need not exist. EM theories emphasise the potential for modern societies to recognise and respond to their environmental impacts by finding new ways of governing environment–economy relations. Key roles are therefore played by governments, markets and technologies, which, under the right conditions, can mean that environmental protection and economic growth become mutually supportive (see Gouldson and Murphy, 1996; Mol *et al.*, 2009). EM theories – with their faith in science and technology, in governments and policy, and in markets and economic growth – tend to be optimistic about the future. Initially many commentators struggled to see how society could deal with a problem on the scale and complexity of climate change, with all of its structural or systemic causes and consequences. However, as Bailey *et al.* (2011) argue, in the past few years the problem of climate change has been rapidly reframed in ways that make it less threatening economically, and more manageable politically. This is a classic manifestation of EM theories; in order to make the climate change problem more tractable, science and technology have been redeployed, the powers of governments and markets reasserted, and environment and economy have once again been portrayed as being mutually supportive rather than mutually antagonistic (Gouldson and Murphy, 1996).

Issues of social learning are at the heart of ecological modernisation – societies need to learn how to detect and understand new environmental issues, how to frame such issues in ways which mean that powerful actors decide to act on them, how to respond to such issues in ways that are technologically, economically and politically viable within the current status quo. EM describes, and thus helps us to understand how some of the major obstacles to change can be negotiated, how problems can be reframed as opportunities, and how more environmentally benign forms of progress can be made possible without radical change to established economic and political systems or practices. EM, in other words, helps us understand how a space can be created within which society can start to learn how to tackle intractable environmental problems such as climate change within existing social contexts – where change and learning are necessary but present opportunities.

Bailey *et al.* (2011) argue that this 'space for feasible action' is defined by the politics of the possible. They argue that EM-like processes depend on the co-option of powerful actors and on approaches that are viable – politically, economically and technologically – within the context of existing institutions and power structures and continued economic growth (Murphy and Gouldson, 2000). EM thus demonstrates what to many is an appealing blend of optimism and pragmatism – rather than wishing that the world was different, it accepts the world as it is and seeks to make progress anyway. Critics, however, argue that such an approach has some major weaknesses.

The first major concern is that the forms of learning that EM seeks to describe and sometimes promote are not deep enough. By focusing on the spaces for feasible action, EM steers the discussion away from those areas that are somehow beyond question in mainstream debates. There are a number of co-conspirators in the dock here – globalisation, liberalisation, economic growth, technological change, materialism, consumerism and so on. In case we needed it, the financial crisis of recent years has revealed a lot about where the priorities really are. In the wake of the crisis, dominant actors have been able to mobilise resources – both political and financial – on an unprecedented scale to support the institutions that facilitate ongoing economic growth. For a brief moment there may have been a discussion on whether these institutions could be challenged or reformed – but that moment was very brief and the mainstream debate ruled that discussion on these issues was inadmissible and very quickly moved on. The same has happened with climate change – those who suggest that an effective response is incompatible with these dominant institutions are quickly marginalised. The space for feasible action on climate change is therefore constrained to those areas that can be accommodated without too much pain by the mainstream actors and institutions. Critics suggest that as a consequence mainstream debates tend to focus on the symptoms rather than the underlying causes of major environmental problems and that as a result they do not really present much of a fundamental solution.

The second major criticism relates closely to the first. It claims that the forms of learning that EM seeks to describe and sometimes promote are unlikely to be quick enough, and that there may be barriers that prevent the deeper learning that we need from happening at all. One EM-type argument that might be put forward is that societies may have to go through a phase where they explore and ultimately experience the limits of the easy options before they are ready to take on the more painful options. This view can be readily attacked. There is no guarantee that the first phase – where some easy options are available – will necessarily lead to a recognition of the limits of such a superficial approach, or to an appetite for deeper change, and even if it might we have not got enough time to wait and see if it actually does. In essence the critical perspective suggests that if radical action will be needed to give us a chance of avoiding dangerous climate change, why spend the next decade learning about the limits of superficial approaches when we can already guess what these might be? We could bypass this phase and get on with the job of developing more fundamental solutions.

Fundamentally, there may never be a resolution to the debate between the proponents and opponents of EM; but in practice many environmentalists waiver between the two positions on an almost daily basis. Do environmentalists in their many guises engage with mainstream debates in the hope that we can have some impact, and risk compromising ourselves and marginalising more radical voices in the process? Or do we refuse to engage in the belief that we will be legitimising a system that is fundamentally incompatible with the goal of sustainability, but decline many opportunities to exert some level of influence in the process? Of course, the world cannot be so simply divided – but in either instance it is useful to understand the processes that construct and constrain the spaces for feasible action on climate change.

The spaces for feasible action on climate change

As is shown in Figure 8.1, the spaces for feasible action can be portrayed as depending on the coincidence of scientific, technological, social, economic and political factors. In essence, this is saying that progress on climate change can be made most readily where it is scientifically justified, technologically possible, economically viable, socially supported and politically acceptable. Measures that develop any of these aspects may be seen to create or constitute the space for feasible action. Within this space, the broader case for change is likely to be strong enough to overcome the self-interested objections of individual opponents to change. Outside this space, there is scope for learning to strengthen the case for action in any particular area and to ensure alignment between the different areas. Spaces for feasible action may therefore be constructed through the different forms of framing and learning that EM seeks to describe, understand and sometimes promote.

Recognition of the importance of this combination of pre-conditions suggests that advocates for action on climate change – or perhaps more accurately advocacy coalitions (Sabatier, 1998) – should adopt a multi-faceted approach that seeks to extend and create alignment between the various factors so that action is made more feasible, particularly at key moments in time when decisions are to be taken. Implicitly at least this is what seemed to be happening in advance of global negotiations on climate agreements at Copenhagen, when many people felt that a combination of pressure, opportunity and alignment between the different pre-conditions for change should have enabled major progress to be

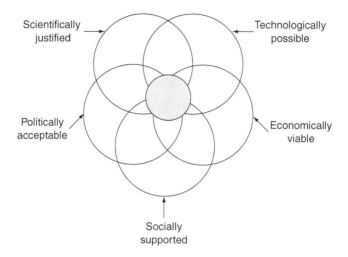

Figure 8.1 Spaces for feasible action on climate change.

made. Of course, one of the key challenges impeding the emergence of more effective global governance mechanisms on climate change has been that these conditions have to be met in a series of different contexts simultaneously – and clearly this hasn't been the case to date. One way around this is to seek to build agreements through a series of smaller and more achievable steps that could collectively represent the 'building blocks' of a more comprehensive global agreement (see Falkner *et al.*, 2010).

By implication, the space for feasible action can be restricted, and progress impeded, if one or more of these pre-conditions for change is not satisfied. Of course it may be that action is still feasible where one or more of the preconditions may be absent if this is more than compensated for elsewhere. For example, if the economic case for action is missing, action may still be possible with sufficient social support or political leadership, etc. However, if all preconditions are needed simultaneously, this creates the prospect of there being various veto points – and this in turn sets the scene for various battlegrounds in climate politics. For example, if popular discourse can be influenced in such a way that it questions the scientific basis for climate change – as it has been in recent years – then this impacts upon various other pre-conditions in the model and the space for feasible action is effectively closed down. Those attacking the integrity of climate science have clearly recognised this – and have called the legitimacy of climate science in its entirety into question because of alleged subjectivities or weaknesses in some particular aspects of the science base. The rigour of core climate science has seen a movement of critique from this to an attack on the personalities of climate scientists and the wider culture and process of science to maintain a window of doubt and space for public critique. The likelihood of the spaces for feasible action being restricted through such tactics is increased where the media – often simplistically – presents climate science as being a series of facts that start to unravel with any admission of uncertainty. Actually – beyond the basic science of global warming – risk and uncertainty are integral aspects of the debate on climate science, particularly when it comes to forecasting the impacts of climate change. Presenting climate science as a series of questions about risk, uncertainty and the balance of evidence may help to make it more resilient to attack.

The macro case in the UK

More detailed empirical support for this conceptualisation of the space for feasible action on climate change is available. Michael Jacobs was special adviser on climate change to Gordon Brown during the period when the UK Climate Change Act was adopted in 2008. He has argued that it was possible to get this ground breaking piece of legislation – which committed the UK to a pathway towards 80 per cent reductions in carbon emissions by 2050 – on to the statute books for a number of reasons (Jacobs, 2010). He argues that it depended upon the collation of an authoritative and trusted evidence base that made a compelling case for action and the creation of an economic discourse which justified action. He also claims that it relied upon the coalescence of a business lobby that makes the case for change and the articulation of clear policy demands by NGOs. He then suggests that it depended upon the presence, for key politicians, of an alignment with their own political identity and narrative. And critically, he argues that – on this issue and in this context – change was only possible at a key moment in time where all of the above factors aligned.

A fuller exploration of the factors proposed by Jacobs (2010) and their presence or absence in the UK during the period leading to the adoption of the 2008 Climate Change Act is warranted. During this period then we can see that there was broad acceptance – at least at the higher levels of government – of the scientific case put forward by the IPCC. There was also a growing acceptance of the viability of certain low-carbon technologies (including nuclear technologies) at a time when significant sections of the energy system needed replacing anyway. This was not (and still is not) a period when a fully functioning energy system with many years of reliable service ahead of it would need to be scrapped and replaced with a new one – major investments have to be made anyway, and this made it easier to make the case for these to focus on the lower carbon options. High-level buy-in to the economic case for action put forward by the Stern Review was also coupled with concerns about the level and volatility of energy prices and with wider concerns about energy security given the decline of North Sea oil and gas reserves and the instability of many alternative sources of oil and gas. These combined with a widespread awareness of the potential for economic benefits through energy efficiency and with significant interest in the potential to develop and export new low-carbon technologies. The lobby for various measures on energy and low carbon was therefore relatively strong and quite diverse.

Critically, the lobby against such measures was not so strong. As an economy dominated by the financial and service sectors, the UK found it easier to act on climate change than countries such as the USA, Australia or Canada where resource- and energy-intensive industries were more powerful. Some of the energy-intensive industries in the UK had also been quite proactive – notably where they had led the development of carbon trading, which ultimately led to the creation of the EU Emissions Trading Scheme. They were therefore relatively familiar with this policy option when it emerged at the EU level – and the adoption of measures at this level ensured that around 60 per cent of UK trade occurred with partners in the EU who in theory at least had to comply with the same standards. This significantly reduced the competitive implications of, and hence levels of opposition to, the policy. The opposition to the EU ETS scheme that did emerge was quickly diffused by starting with relatively unthreatening standards and by handing out free carbon permits (and thereby awarding significant windfall profits) to some energy-intensive sectors.

Some key NGOs also grouped together to present a powerful argument for change. Friends of the Earth in particular spent a lot of time developing and promoting detailed and plausible proposals that were very close to what later became the Climate Change Act. When coupled with a government that was not afraid to intervene, all of this made it politically possible to act. The one aspect that was perhaps lacking was active public support for ambitious low-carbon targets. In the absence of this, a level of political entrepreneurship ensured that ambitious carbon budgets and targets were almost 'flown in under the radar' and then locked into law. More recently, attempts have been made to build public support for the targets and the transition in the early phases of the transition when relatively pain-free progress can be made. It seems likely that this support will become more and more important once the relatively easy options have been exploited and as the subsequent phases become more challenging, whether technologically, economically or politically.

This raises an important question about the relative importance of all of the factors that need to be aligned to create a space for feasible action. Is action on climate change possible in contexts where there is a comparative absence of social support, or where support from some is countered by opposition from others? Similarly, can there be action on climate change when the economic case is not clear, or where support from the winners (i.e. the renewable industries) is outweighed by the opposition from the losers (i.e. the fossil fuel or energy-intensive industries)? Clearly, insights from the realms of political economy are relevant here – and further research is needed. It seems likely though that action can most easily be achieved when all of the factors are aligned, but that it can still be achieved under some circumstances (i.e. with strong political leadership at the national level, or with agreement and common action at the global level) even if some elements are absent or are not aligned.

The micro case in the UK

The preceding narrative about the spaces for feasible action on climate change at the macro level is mirrored in many ways at the micro level. Focusing on the business sector, we can see that for many firms climate change emerged and rose quite rapidly up the list of priorities in the period from 2006. At least for some of the larger and higher profile firms, a space for feasible action on climate change seems to have emerged. This raises significant questions about why this space emerged, how big it is, what can be done within it and how long it will last. This in turn begs a question about the range of factors that govern corporate behaviour on climate change.

Of course, in some settings a primary factor that drives corporate behaviour on climate change is government policy. This is certainly the case for the energy-intensive industries where the EU ETS is a significant influence, and in the UK, this level of influence has been extended through the 2008 Climate Change Act to include many more significant energy users through the introduction of the Carbon Reduction Commitment (CRC). But even for these sectors it is likely that behaviour is driven more by markets than by governments. The levels and the volatility of energy prices are probably more of an influence on corporate energy use and carbon footprints than government policy. This can be of some comfort to governments that are having (or in some cases are happy) to accept that their powers are limited – we are never far away from a neoliberal logic which says that markets can look after the climate better than government, especially in an era of peak oil and geopolitical instability. More and more frequently, governments seem to be working with rather than against the grain of markets to promote change – and it seems increasingly difficult for them to work against markets. This is leading to some innovations in governance – exemplified in the UK by ongoing discussions about a Green Investment Bank to provide additional funding for large-scale low-carbon projects, or about a Green Deal to provide similar funding for smaller scale measures in homes and smaller businesses.

Alongside the influence of governments and markets, a range of other actors seem to be playing a supporting role in promoting corporate action on climate change. Some institutional investors have taken an interest in climate change and issues related to energy – often asking the companies they invest in to show that they have the capacity to manage energy and any associated risks (i.e. from future regulation). While for many firms climate change and energy use are not seen as 'material' issues that have a significant effect on share price or profitability, their ability to manage energy efficiently may reveal something about their ability to manage other aspects of their cost base effectively and hence there is some pressure to do so. To some degree the pressure is accentuated through different reporting or disclosure schemes. In recent years, the NGO-led Carbon Disclosure Programme (CDP) has grown significantly. While its aim is to collect data and to increase transparency on carbon-related issues, even with the data it provides it is still virtually impossible to reliably assess the carbon or energy performance of a firm either over time or to compare its performance with that of another company. As a result the competitive pressures that emerge from mainstream financial accounting and reporting have yet to bite in the same way for carbon or energy. It is easy for companies to voluntarily disclose information on their carbon or energy performance if they know that the data do not allow direct comparisons to be made either over time or between companies. In essence they can disclose but still be unaccountable. The potential influence of a significant governance instrument is therefore lost.

More tangible pressure can come from other areas – for example, many larger firms are putting pressure on their suppliers to meet higher standards on carbon and energy. Often these pressures are primarily motivated by cost, and their influence stops when the opportunities for cost-effective ways of reducing energy and carbon dry up. More broadly, companies are being presented with new opportunities by various technology providers, energy companies and energy service companies. Concerns about climate change also feature in surveys of customer priorities (although not often in the top five factors), and media and NGO campaigns can occasionally impact directly upon some high-profile companies, as has been the case in the UK where campaigners have stopped or slowed down the construction of runways or coal-fired power stations.

What is interesting is that it can be difficult – particularly for those sectors of the economy that are comparatively unregulated – to trace the influence of any one factor on the business case for action on climate change. Instead it seems that climate change may have made it on to the agenda for some large companies owing to the combined effects of a range of different and sometimes diffuse measures. The publication of the Stern Review in 2006 and the IPCC's *Fourth Assessment Report* in 2007, and the adoption in the UK of the Climate Change Act, 2008 all provided momentum for discussions that led into Copenhagen at the end of 2009, and conspired to convince the leaders of some companies that climate change was a big enough issue that would be around for a long enough time to demand a response. When coupled with a period of high and volatile energy prices and a sense that it was technologically and economically possible to do something about it, some companies decided that they should indeed do that. Again, therefore, the space for feasible action on climate change has depended upon the coincidence of a range of different factors.

A critically important issue here relates to the economic case for action on climate change. Corporations have been happy enough to accept targets and regulations and to take various voluntary measures because in this early stage of the transition to a low-carbon economy there are often economic returns from doing so. The climate-related targets included within Marks and Spencer's Plan A campaign are an example of this. The space for feasible action depends on this being the case. But what happens once the easy options have been exploited? Here we see some of the issues about EM being played out in miniature. Will companies learn from the early phases of transition where the changes required are reasonably superficial, so that when the challenges become more significant (as they surely will if we are to achieve 80 per cent reductions in carbon emissions) they are better able to cope? Or will they go along with business as usual so long as it is economically attractive to do so, but then gradually withdraw their active support and drift towards active opposition as the changes required become more challenging? As yet there is very little evidence that more than a tiny minority of larger companies have thought through what a radically decarbonised business or economy might look like – and there is very little evidence to suggest that many would be willing to rethink their business models to secure deeper carbon cuts (see Gouldson *et al.*, 2011, forthcoming). More frequently we see that efficiency improvements are just enough to offset the increased energy use and carbon emissions that occur as a result of business growth.

Conclusions

So what are the implications of the preceding debate for climate policy and for the prospects for a transition to a low-carbon economy and society?

It is clear that the challenges preventing the realisation of a global agreement on climate change are formidable. In essence, a space for feasible action on climate change has to open up at the same time in all of the major economic powers if we are to secure agreement. This seems unlikely – even with the innovations and learning that are at the heart of EM, spaces for feasible action may be small and temporary: they open and close in different contexts at different times. Waiting for such moments may not be appropriate given the scale and proximity of likely dangerous climate change and continuing resistance to low-carbon pathways. An alternative, 'building blocks' approach seems a more plausible way forward (Falkner *et al.*, 2010). The dynamics of the political negotiations could change significantly if the front runners of this process could protect themselves from the free riders in some way, for example, through border adjustments on embedded carbon.

In particular contexts, though, there is of course scope for learning. For governments, climate change has emerged as an issue that requires significant levels of action at a time when their ability to act, and sometimes also their willingness to act, have been restricted through the processes of globalisation and liberalisation and all of the attendant concerns about competitiveness (see Reinaud, 2008). However, in some settings we have seen an ambitious reassertion of the powers of the state on climate change – the setting of targets for decarbonisation that could radically transform our societies and economies in the coming years (i.e. through the UK's Climate Change Act) can certainly be seen in this way.

In some contexts this has been made possible through innovations in policy and governance. This is closely associated with the much-vaunted transition from the provider or controller state to the facilitator or enabler state (see Gouldson and Bebbington, 2007). This involves governments creating the conditions within which private or civic actors can deliver public interest objectives. Where governments do not have the political or economic power to deliver the transition on their own, by harnessing the powers of markets and civil society they are clearly able to do more. But in doing so how much power are they ceding, and does this mean they can only do as much as markets or society at large are willing to do? High-profile examples of public opposition to wind farms or fuel tax protests highlight the power of aspects of civil society to mobilise against unpopular decisions relating to climate change. Industrial lobbying over concerns about over-regulation and its impacts on competitiveness is often less high profile but it certainly impacts upon government decision-making. The powers of government – even (or perhaps especially) when it harnesses the powers of markets and civil society – can be limited in various ways. This concern resonates with deeper worries about the shifting nature of the social contact in societies brought about by climate change impacts and policy described also elsewhere in this book.

So what are the prospects for governance beyond the state? There are numerous corporate and civic initiatives that seek to promote different forms of action on environment and climate change. Of course there can be various motivations to participation in these. While there may be a level of altruism, or a sense that there is an important but unquantifiable strategic benefit, in general it is unlikely that a critical mass of organisations or individuals would voluntarily commit to initiatives that might damage their economic interests. The space for feasible action may again be constrained by the extent of the economic case for action. However, some level of action beyond the boundaries of such an economic case may be feasible if non-state actors are able to exert some amount of coercive power. Although in practice this often seems unlikely, in theory this could happen – NGOs, communities, the media and others could draw upon their moral authority to legitimise or delegitimise different forms of action, particularly where they are empowered to do so through different forms of access to information. League tables and naming-and-shaming campaigns can certainly be drivers of change, and corporations (especially those with significant brands to protect) certainly pay attention to the reputational risks that they are exposed to. In addition, there may be opportunities for private or civic actors to 'nudge' different actors so that they behave in more environmentally friendly or less carbon-intensive ways (Thaler and Sunstein, 2008). But in broad terms the potential of non-state actors to drive change seems to be constrained by the extent of the economic case for action. If the economic case for action dries up, then the space for feasible action will either need to be closed down or progress within this space will be harder to achieve.

All of this suggests that the 'governance turn' that has been taken in the UK and the EU and elsewhere in recent years may well be able to drive change in the early and relatively pain-free stages of transition. But it seems unlikely that such governance mechanisms will be strong enough or durable enough to drive change in the subsequent phases of the transition where change becomes more challenging, whether socially, technologically, economically or politically. The proponents of EM theories might suggest that we can learn and innovate in ways that will delay the point at which we encounter this phase of the transition. But others might say that in the end the transition to an 80 per cent decarbonised world will have to rely less on consensus and more on government action and leadership.

References

Bailey, I., Gouldson, A. and Newell, P. (2011) Ecological modernisation and the governance of carbon: a critical analysis. *Antipode*, online advance publication DOI: 10.1111/j.1467–8330.2011.00880.x.

Enqvist, P., Naucler, T. and Rosander, J. (2007) A cost curve for greenhouse gas reduction. *Mckinsey Quarterly* 1.

Falkner, R., Stephan, H. and Vogler, J. (2010) *International Climate Policy after Copenhagen: Towards a 'Building Blocks' Approach*, CCCEP Working Paper 25, Centre for Climate Change Economics and Policy (www.cccep.ac.uk).

Giddens, A. (2002) *Runaway World: How Globalisation is Reshaping Our Lives* (2nd edn), London: Profile Books.

Gouldson, A. and Bebbington, J. (2007) Corporations and the governance of environmental risks. *Environment and Planning C* 25 (1): 4–20.

Gouldson, A. and Murphy, J. (1996) Ecological Modernisation and the European Union, *Geoforum*, 27 (1), 11–21.

Gouldson, A., Hills, P. and Welford, R. (2008) Ecological modernisation and policy learning in Hong Kong. *Geoforum* 39: 319–330.

Gouldson, A. *et al.* (2011) Business perspectives on the transition to a low carbon economy. *Business Briefing*, Centre for Low Carbon Futures/Centre for Climate Change Economics and Policy.

Hulme, M. (2009) *Why We Disagree About Climate Change: Understanding Controversy, Inaction and Opportunity*, Cambridge: Cambridge University Press.

IPCC (1990) *The First Assessment Report*, Inter-Governmental Panel on Climate Change.

IPCC (2007) *The Fourth Assessment Report*, Inter-Governmental Panel on Climate Change.

Jacobs, M. (2010) 'The Climate Challenge: Creating Political Will', address to Centre for Climate Change Economics and Policy Conference, 'Grand Challenges in the Transition Towards a Low Carbon, Climate Resilient Society', London School of Economics, 16 September.

Mol, A., Sonnenfeld, D. and Spaargaren, G. (eds) (2009) *The Ecological Modernization Reader: Environmental Reform in Theory and Practice*, London: Routledge.

Murphy, J. and Gouldson, A. (2000) Integrating environment and economy through ecological modernisation? An assessment of the impact of environmental policy on industrial innovation. *Geoforum* 31: 33–44.

Paavola, J., Gouldson, A. and Kluvankova-Oravska, T. (2009) Institutions, ecosystems and the interplay of actors, scales, frameworks and regimes in the governance of biodiversity. *Environmental Policy and Governance* 19 (3): 148–158.

Pacala, S. and Socolow, R. (2004) Stabilization wedges: solving the climate problem for the next 50 years with current technologies. *Science* 305 (5686): 968–972.

Reinaud, J. (2008) *Issues Behind Competitiveness and Carbon Leakage*, IEA Information Paper, International Energy Agency and Organisation for Economic Cooperation and Development.

Sabatier, P. (1998) The Advocacy Coalition framework: revisions and relevance for Europe. *Journal of European Public Policy* 5 (1): 93–130.

Stern, N. (2006) *The Economics of Climate Change*, Cambridge: Cambridge University Press.

Thaler, R. and Sunstein, C. (2008) *Nudge: Improving Health, Wealth and Happiness*, New Haven, CT: Yale University Press.

Part III

Beyond capitalism

Critical theory and de-growth

9 Climate change, 'the cancer stage of capitalism' and the return of limits to growth

Towards a political economy of sustainability

John Barry

There are no ... limits to the carrying capacity of the earth that are likely to bind any time in the foreseeable future. There isn't a risk of an apocalypse due to global warming or anything else. The idea that we should put limits on growth because of some natural limit, is a profound error and one that, were it ever to prove influential, would have staggering social costs.

Lawrence H. Summers, Chief Economist World Bank, 1991

Introduction

The critique of conventional economic growth has been a long-standing position of green thinking and radical conceptions of sustainability. Questioning economic growth is one of the most radical challenges to lay before capitalism, given it is a political economic system structurally 'locked into' continuous economic growth and accumulation. Indeed, one could suggest that any plausibly 'green' and radical conception of political economy begins from and articulates a 'limits to growth' perspective (Barry, 1999). Now while there are many debates as to understandings and measurements of 'economic growth', a 'post-growth' economy is one that has featured prominently within green political and economic discourse, most usually associated with the environmental and political benefits of a less growth-orientated socio-economic system. This 'post-growth' insight establishes the first framing issue of this chapter which is to reject conventional technologically optimistic ideas of 'decoupling' resource, energy and pollution from orthodox economic growth (Jackson, 2009), most commonly encapsulated in the dominant official state and corporate reading of sustainable development as 'ecological modernisation' (Barry, 2005), 'green growth' (UNEP, 2011), 'green' or 'natural capitalism' (Hawken *et al.*, 1999). That such technologically based wishful thinking informs most government policies in relation to climate change and the ecological crisis not only demonstrates the structural imperative to maintain 'business as usual' in terms of orthodox economic growth and globalised capitalism, but also the ideological power of this form of economic thinking, even in the face of mounting scientific evidence that the growth imperative itself is the major cause of climate change, biodiversity loss, deforestation and so on.

Second, as this chapter seeks to explore, the discourse of 'resilience' – unlike that of conventional 'sustainable development' – moves us away from 'win–win' 'business-as-usual scenarios' in relation to the compatibility of orthodox economic growth and social and ecological sustainability, and forces some hard questions to be confronted. Two of these hard questions are (1) issues of major structural transformation in social, economy and political relations and institutions cannot be avoided in the transition to a sustainable society, including the transcending of capitalism, and (2) in this way resilience, as understood in this chapter, is to be viewed not merely as a 'reactive' 'coping mechanism' to enable a person, community or society *to return to some status quo ante*. Of course it can have this politically conservative meaning, especially if one simply takes a narrow ecological account of resilience, and/or wishes to ensure the continuing 'functioning' of the existing, capitalist economy and system. However, drawing on a variety of approaches to resilience ranging from permaculture (Holmgren, 2009), the Transition Movement (Hopkins, 2008), and complex adaptive systems (Folke, 2006). As Folke puts it, 'The concept of resilience shifts policies from those that aspire to control change in systems assumed to be stable, to managing the capacity of social-ecological systems to cope with, adapt to, *and shape change*' (Folke, 2006: 254; emphasis added). Or as Ted Trainer notes from a Transition Movement view, giving it a clear radical and anti-capitalist character:

> It is not oil that sets your greatest insecurity; *it is the global economy*. It doesn't need your town.... It will only deliver to you whatever benefits trickle down from the ventures which maximise corporate profits. It loots the Third World to stock your supermarket shelves.... In the coming time of scarcity it will not look after you. *You will only escape that fate if you build a radically new economy in your region, and run it to provide for the people who live there.*
>
> (Trainer, 2009: 14; emphasis added)

Thus, transformation, evolution, change and transition (at multiple scales) are central to how resilience is understood in this chapter. That is, to make the transition to becoming a resilient community or society is, in large part, based on a political choice to live in a *different* type of community (as in the Transition Movement) or society (as in eco-socialist or green political perspectives) than the current unsustainable, unequal and non-resilient one. Here, one could compare this focus on resilience and green political economy to recent strategies for versions of a 'Green New Deal' as denoting a greening 'business as usual'. Green New Deal proposals in the main seek through green Keynesian stimulus packages to respond to the current global economic recession through government investment in large-scale low-carbon infrastructural projects, boosting employment and providing a 'step change' to a low-carbon economy (Green New Deal Group, 2008; Luke, 2009; Barry, 2010). Such proposals (which promise the compatibility of kick-starting economic growth/recovery, international competitiveness and the decarbonisation/greening of the economy) are particularly associated with economic and public policy proposals for addressing climate change. In particular Nicholas Stern, coordinator of the influential UK Treasury report into the economics of climate change in 2006 (Stern, 2006), is a prominent proponent of government investments in green energy and economic infrastructure to address both the economic and climate crises (Bowen *et al.*, 2009).

Third, the issue of economic growth under capitalism is at the heart of how we should understand the underlying causes of climate change (and its flip side – peak oil and the decline in cheap and secure sources of carbon energy) as opposed to simply focusing on its effects. In terms of the dominant mitigation and adaptation approaches to climate change, while economic growth does have significant impacts on adaptation policies, this chapter, in outlining a green political economy perspective, will focus on how effective and meaningful approaches to reducing the causes of climate change need to focus on questioning orthodox economic growth. A green political economy perspective is one which politics, struggle and power relations are central to any analysis of the economy and how 'economics' is understood, but which also sees as central the dynamic, mediated and metabolic (energy/resource/pollution) relationship between the human economy and the non-human world (Scott Cato, 2009). For example, and in direct contradiction of the orthodox 'green new deal' approach, at the recent Cancún climate talks in December 2010, Kevin Anderson, director of the Tyndall Centre for Climate Change, proposed that avoiding dangerous climate change was not compatible with continued economic growth in the developed world. As he notes in a co-authored article, 'only the global economic slump has had any significant impact in reversing the trend of rising emissions' (Anderson and Bows, 2011: 38). This paper goes on to state that 'reductions in emissions of 3–4% per year are not compatible with economic growth … avoiding dangerous (and even extremely dangerous) climate change is no longer compatible with economic prosperity' (ibid.: 40). The paper concludes that

> the logic of such studies suggest (extremely) dangerous climate change can only be avoided if economic growth is exchanged, at least temporarily, for a period of planned austerity within [rich] nations and a rapid transition away from fossil-fuelled development within [poorer] nations.

> (ibid.: 41)

In keeping with long-standing green 'limits to growth' concerns (as well as more contemporary developments such as the 'degrowth' or 'decroissance' movement (Degrowth Declaration, 2010)), Anderson's scientifically informed position suggests that *planned economic contraction in the developed or minority world* is inevitable if we are to avoid dangerous climate change.

However, while there is a weight of evidence we can cite to question economic growth and outline reasons for adopting a 'limits to growth' position as the starting point for any informed political analysis of climate change, there is less research and debate around the alternative to economic growth under capitalism. But what replaces conventional economic growth? Stagnation? Collapse? Regress? Is a non-growth orientated capitalism possible or desirable? On first gloss the prospect of a non- or post-growth economy not only sounds odd and unfamiliar but also negative and perhaps even dangerous and harmful. In a commonsense context 'growth' is normally something we perceive as 'good', something positive to be promoted, since it indicates maturity, development, positive movement. Children grow, plants grow, etc. and growth is therefore normally and uncontroversially viewed as something to be promoted and encouraged. But also in technical terms growth is not simply 'good' for the capitalist economy but indeed a functional (i.e. system) requirement. As Jackson points out, 'in a growth-based economy, growth is functional for stability. The capitalist model has no easy route to a steady state position. *Its natural dynamics push it towards one of two states: expansion or collapse*' (2009: 64; emphasis added). In other words, within a capitalist growth-orientated economy, a shift away from growth leads to recession, socio-economic instability, job losses, investment uncertainty and a decline in living standards, etc.

This chapter will seek to both criticise the dominant model of political economy (namely neoclassical economics) under capitalism and which informs public policy, and its commitment (addiction would also not be inappropriate) to 'economic growth'. It will also outline a green political economic alternative to economic growth based around the notion of 'economic security' which, it will be argued, represents an attractive and radical way of addressing climate change, social and global inequality and delivering well-being to people. In so questioning economic growth, we are also questioning capitalism and therefore in analysing climate change (and resilience) this chapter suggests we need a clearer 'post-capitalist' political economy perspective, one that could (potentially) bring the ecological and labour/socialist movements together (though the latter's embracing of a 'post-growth' analysis is far from certain). In this way, what this chapter suggests might have practical, strategic as well as analytical merit.

Economic growth and the 'cancer stage' of capitalism

John McMurtry criticises economic growth in a provocative and indeed troubling manner. For him, economic growth under capitalism is understood as 'growth for growth's sake' and is therefore for him a form of 'cancerous growth', the logic of which constitutes a clear and present threat to 'life' (i.e. both human and non-human) on the planet (McMurtry, 1996). Cancer, understood as denoting the threshold beyond which cell growth becomes unhealthy and threatening the body, may be understood as 'growth for growth's sake'. In the same way, economic growth as 'growth for its own sake' is cancerous in that it is not growth orientated towards some of the end, such as improving human well-being or the habitats of the non-human world. Rather economic growth under capitalism, as Marx, and those who draw inspiration from his political economic analysis, demonstrated, is orientated towards capital accumulation, not improving the life of humans or the non-human world.

McMurtry's argument is that capitalist economic growth is destroying life and the life-supporting mechanisms of both people and planet. Echoing early critical theorists such as Herbert Marcuse, McMurtry foregrounds how the 'normal' and 'mundane' operation of the capitalist economy undermines 'life' itself (human and non-human) as well as 'the lifeworld'. For Marcuse, drawing on Freud's insights, capitalism's relentless drives for 'more', 'exponential growth', speed, mobility, etc. are all expressions of a 'death instinct' (Thanatos) preying on or being parasitic upon the 'life instinct' (Eros). The important point here is the notion that beyond a threshold, economic growth becomes unhealthy, unsustainable and therefore something to be viewed as potentially harmful rather than as something to be actively sought after and uncritically promoted as a self-evident 'good'. In the words of the ecological economist Herman Daly, there is such a thing as 'uneconomic growth' and 'illith' (Daly, 2003). It is not that economic growth is to be abandoned but rather to be viewed as a process to be consciously and politically monitored and regulated, and which has a threshold, an end point, rather than viewed unreflectively as something that can be simply 'left on automatic' to a self-regulating economic system. In other words, a green critique of economic growth sees an unthinking commitment to its *infinite continuation as a permanent objective of public policy* as both socially and ecologically irrational.

This is important in terms of the common misunderstanding of those critical of economic growth as proposing that countries (mostly in the Global South) that can benefit from economic growth (and trade) should be denied that opportunity. Following one of the first modern thinkers to propose a 'post-growth' economy, namely, John Stuart Mill (Barry, 2007) and his advocacy of a 'stationary state' for the economy, green political economists are of the view that economic growth should be 'redistributed' from the 'overdeveloped' minority world in the North/West to the majority world in the Global South. It is for this reason that debates about a 'post-growth' economy are squarely orientated towards the (over)developed world. They should be viewed within the context of global (as well as national, regional and citizen or per capita) arguments for less unequal distribution of opportunities, income, work (not to be confused with formally paid 'employment'), and the meeting of human needs. And ultimately, as this chapter will argue, a post-growth political economy perspective is concerned about the more equal distribution of economic security and well-being, in opposition to the unequal distribution of conventional economic growth under capitalism.

The grammar of public policy: the impact of economics upon life, lives and livelihoods

The reasons why the stakes are so high with regard to the dangers posed by one paradigm dominating our thinking about the economy should be obvious: if economics were simply on a par with cultural studies or history, debates about it would not have the same political character, since these debates and controversies would be seen as 'internal' to the academy and the discipline with few real-world implications. However, given that how we conceptualise economics frames decision-making about the distribution of resources, tax policy, reform of health, education, trade, aids and informs our approach to such pressing issues as climate change means that the battle over how economics is understood and what the economic objectives of a society are is at the heart of the politics of climate change policy.

Perhaps the greatest success of the neoclassical orthodoxy lies in its being the *grammar* of policy-making of capitalism orientated towards producing infinite economic growth as a structural necessity of the political economic system. That is, neoclassical economics has embedded itself so successfully within decision-making (or co-evolved with it) that it not only acts as 'gatekeeper' and agenda-setter but it also determines the language and way in which those wishing to influence or have input into public policy-making must express their argument. From a Foucauldian perspective, neoclassical economics is a knowledge/power discourse which shapes the way we think, act and decide public policy. Neoclassical economics becomes a 'truth regime' and constitutes the very 'rules of the game' in the same way as grammar is the rules for the correct use of language. Thus those who either do not know or refuse to accept this particular grammar (such as non-economic arguments for environmental preservation or those economic perspectives critical of the neoclassical framework) are at a severe disadvantage in trying to influence environmental policy-making within the current institutional and power/knowledge framework.

A key feature of the power of neoclassical economics is its 'success' in terms of 'delivering the goods' for a period of time, creating a 'perpetual growth machine' supported by those who have benefited from orthodox economic growth under neoliberal globalisation. The reality is that however much one can and ought to criticise this mode of thinking and policy-making about the economy, it did lower prices for food and basic commodities (of course passing the full cost on to workers in other parts of the world on low wages, or to local environments and ecosystems, other forms of non-human life, or the global climate system) which allowed more people in the developed world to enjoy lifestyles once the preserve of the wealthy. However, when the full 'on' as opposed to 'off-balance' sheet is reviewed, it is clear that this perpetual growth machine was and is maintained at tremendous environmental and human cost and was literally fuelled by (now declining) carbon energy. With ecological chickens coming home to roost and the beginning of the end of the carbon energy age, we need modes of thinking about the economy which are 'fit for purpose' for the challenges and opportunities of our current predicament. One such alternative to neoclassical economics, namely green political economy, is outlined below.

The battle of ideas over the economy

The success of a body of knowledge is not only to be found in whether it 'trumps' other potential forms of knowledge in the pages of an academic journal or conference panel, *or indeed in replacing one 'paradigm' with another*, but also in whether it exists, persists and has support in the 'real world'. Of course one need not be related to the other. As we can see in relation to neoclassical economics, even though this body of knowledge not only failed to predict the current global economic crisis but was also the main cause of the policies which precipitated it, it is still neoclassical economists who continue to advise governments, business models have not by and large been transformed, and ordinary citizens have been offered no alternative to the neoclassical orthodoxy, as may be seen in the media coverage of the crisis. Much like the 'Achilles lance' view of economic growth – namely it can heal the wounds (environmental and social) it inflicts – so likewise with the neoclassical orthodoxy. Thus we find that the solution to the problems caused by neoclassical economics and growth orientated capitalism is … more neoclassical economics prescriptions and a continuation of growth-orientated capitalism (with a 'Green New Deal' twist).

As Scott Cato perceptively notes, part of the reason for the absence of green economics from the academy has to do with the sense that 'academic debate around economics and, some would argue, the role of the university itself, has been captured by the globalized economic system, whose dominance is a threat to the environment' (Scott Cato, 2008: 6). Conversely, viewed as an ideological battle, debates about economics cannot be confined to the academic sphere but must also be analysed and located in everyday life and individual life histories (as Manuel-Navarrete suggests in Chapter 10, this volume), the public sphere, popular culture, the media, elections, party manifestos, protests, campaigns, commonsense perceptions about the economy, to public policy-making and legislation.

This is of course an old Marxist insight – namely that debates and conceptions about the economy and economics are, in part, about bolstering or unsettling power relations within society, a battle for ideological hegemony. In the case of neoclassical economics, its ideological hegemony translates not simply into political power in determining state policies, for example, but is equally a form of cultural hegemony informing how we 'commonsensically' think about, assess and evaluate the economy and economic issues. Perhaps the most vivid expression of an ideological position that has achieved this pre-eminent position is that it neither presents itself as an ideological position nor is perceived as such by others, but rather is viewed as 'commonsense' or 'normal'. Once a particular way of conceptualising and thinking about the economy is widely shared and commonsensical, alternative modes of thinking about the economy are by default 'nonsensical', and indeed this has and continues to be the most common reaction to non-neoclassical economic perspective – green or other.

In other words, a category mistake has been made (the confusion of 'capitalism' with the 'economy') which despite the normatively charged debates within the academy continues to persist in the real world. Much the same of course occurs with other central terms in modern societies such as 'democracy' being reduced to 'liberal' or representative democracy, other conceptions of democracy are marginalised and neglected or simply deemed 'abnormal' or otherwise suspect.

From economic growth to economic security

So what can replace 'economic growth'? While there are many potential contenders for this that have been canvassed over the past 150 years – from John Stuart Mill's 'stationary state' (Barry, 2007) to more recent work on 'quality of life', 'well-being' and 'prosperity without growth' (Jackson, 2009), all share key components and have a large degree of overlap – the one I wish to explore here is the notion of 'economic security'. I begin from a major report by the International Labor Organization, *Economic Security for a Better World*, published in 2004, which found that 'economic security' coupled with democracy and equality were key determinants of well-being and social stability. This study developed an Economic Security Index. According to the report:

> *Economic security* is composed of *basic social security*, defined by access to basic needs infrastructure pertaining to health, education, dwelling, information, and social protection, as well as *work-related security* ... two are essential for basic security: income security and voice representation security.
>
> (ILO, 2004: 1)

The report also found that:

> People in countries that provide citizens with a high level of economic security have a higher level of happiness on average, as measured by surveys of national levels of life-satisfaction and happiness.... *The most important determinant of national happiness is not income level – there is a positive association, but rising income seems to have little effect as wealthy countries grow more wealthier. Rather the key factor is the extent of income security, measured in terms of income protection and a low degree of income inequality.*
>
> (ILO, 2004: 1; emphasis added)

Such findings give empirical support to long-standing green arguments stressing the need for policies to lower socio-economic inequality, enhance individual and collective socio-economic security and increase well-being and quality of life, rather than conventionally measured economic growth, rising personal income levels or orthodox measures of wealth and prosperity. The link between economic security, well-being and equality will be explored in more detail below. Thus the ILO report finds that welfare state provision of social security for citizens is an important determinant of high economic security. The ILO report also points out that insecurity is generated by patterns of economic globalisation which produce endemic or structural insecurity in terms of employment, social welfare and income due to countries 'racing to the bottom' in terms of lowering worker protection and welfare provision, or how 'footloose' multinational capital and the demands of international economic competitiveness by their very nature undermine economic security as companies can always relocate in search of higher profits, lower wages, etc.

It is also important to note other observed links between (in)security, economic growth and well-being. One of the principal psychological and cultural determinants of excessive consumption has been found to be feelings of personal insecurity and vulnerability – whether about one's body shape, sensitivity to peer judgements, or externally generated and reinforced views of self–other relations that undermine personal or other forms of security and self-esteem (Chaplin and John, 2007). Giddens, in a slightly different vein, has written persuasively about the way in which modernity can undermine what he calls 'ontological security' (Giddens, 1994: 79), but does not connect this with patterns of 'defensive consumption' – those forms of consumption which do not add to quality of life, but are forced upon the individual as a necessary means for them to simply protect their existing material standard of living (i.e. to run to stand still as it were). Ego insecurity may therefore be viewed as a main cause of consumerism and materialism, which lead to diminished well-being. According to psychologist Tim Kasser's work, 'Materialistic people, from children to pensioners, are less satisfied with life, lack vitality, and suffer more anxiety, depression and addiction problems. Materialistic values make people more anti-social, less empathic, more competitive and less cooperative' (Kasser, 2008: 92). He points out that individuals when faced with insecurity or pain (psychological or physical) in the USA increasingly turn to money and possessions as a way of coping with distress rather than seeking comfort and support in social interaction and community or family relationships.

It turns out that the conditions for human flourishing are intimately connected to the conditions for transformed and resilient communities. While there are clear material/resource inputs needed for creating resilient communities and meeting human needs, much of what is required for human flourishing and resilience is non-material, once a decent standard of living has been attained, and are largely to do with relations, social and communicative interaction and central to which is less unequal power relations within and between communities and individuals (Cutter, 2006; Wilkinson and Pickett, 2009). The upshot of this is that human flourishing and providing, and equitably distributing economic security, can be achieved at much lower ecological and resource throughput. That is, economic contraction/degrowth/post-growth (particularly if we remember that we in the minority world are starting this 'descent' from a high level of material wealth, though inequitably distributed) does not necessarily mean lower levels of human well-being. However, a key issue, and one which space does not permit a fuller exploration here, is that such an economic contraction, and the associated 'retreat from fossil fuels', must be planned. Thus, it does seem that what a 'macro-economics of sustainability' in a post-growth context calls for is *economic planning* and not just more *economic regulation*. This does seem to open up a fruitful area of cooperation and debate between the green and labour movements.

Of course there are those for whom economic insecurity is viewed in a positive manner and indeed held as a necessary feature of modern capitalism. Joseph Schumpeter, for example, famously noted capitalism's 'creative destruction' by which he meant that 'This process of Creative Destruction is the essential fact about capitalism. It is what capitalism consists in and what every capitalist concern has got to live in' (Schumpeter, 1975: 82). For him, and indeed for other mainstream economists, a key feature of capitalism's productive dynamism is its inherent instability and its capacity to generate insecurity as a way of spurring innovation and entrepreneurialism, all with the aim of course of increasing productively, profitability, and therefore economic growth and capital accumulation. Thus, arguments for economic security run counter to this vision, both in questioning the means (creative destruction and insecurity) and the ends (orthodox economic growth as measured by GDP). However, this does not necessarily mean (as critics are wont to point out) that a focus on economic security as a main objective of macro-economic policy puts an end to entrepreneurialism or innovation. It is a fair comment to make and one which any alternative to our current economic growth-focused model needs to take seriously is how to ensure that stagnation and regress is not the outcome of a post-growth economy. However, there are solid reasons for thinking that economic development 'comes from innovation, from consuming different things, rather than more of the same things' (Wilkinson and Pickett, 2009: 221).

Beyond maximisation and efficiency: sufficiency and frugality and the macro-economics of sustainability

Some notion of sufficiency would have to be harnessed to our understanding of economic security, if the concept is to adequately service our needs for both social well-being and socio-ecological resilience. Herman Daly while welcoming the idea of integrating the principle of sufficiency also signals the difficulty of doing so. As he puts it, 'It will be very difficult to define sufficiency and build the concept [of sufficiency] into economic theory and practice. *But I think it will prove far more difficult to continue to operate [as if] there is no such thing as enough*' (Daly, 1993: 360–361; emphasis added). Sufficiency principles (as opposed to mere efficiency) such as those of restraint, respite, precaution, have the virtue of partially resurrecting well-established notions like moderation and thrift, ideas that have never completely disappeared and will be in need as guides to action in a sustainable, resilient economy. An important point here is raised by Astyk in correcting the common misperception of thrift and frugality being negatively viewed. As she rightly puts it,

> Thrift is not the opposite of generosity, the closed fist that holds on to what you have, but the enabler of generosity. A frugal life that does not waste and cares for what you have is what enables you to give away, to share, to open your hands and pour forth what you have preserved.

> (Astyk, 2008: 208)

Frugality as a principle of economic thinking and acting has perhaps been regarded as a form of private 'accumulation strategy' due to the unfortunate association between thrift, saving and frugality with Scrooge-like, capitalistic motivations (i.e. accumulation). In fact, as outlined below, investment and saving, based on principles such as frugality and thrift, mean that the latter principles represent the antidote to debt-based private consumption and, as Astyk and others note, can move us in the direction of a more solidaristic, sharing (and planned) economy.

The concept of sufficiency also translates to the relationship between production and reproduction, 'employment' and 'work'. Once we begin to see that the point of public policy or how we should judge the success or not of the economy is how it provides meaningful work which contributes directly to social well-being, has forms of provisioning which meet people's needs, and balances this with formally paid employment and the needs of the formal economy, we are entering a very different economic worldview. While of course bound to attract criticism, and not for one moment denying or minimising the devastating effects of losing one's job, in our current economy, I do think the time has come to seriously revisit Ivan Illich's wonderfully provocative notion of *The Right to Useful Unemployment* (Illich, 1978). In short, we need to ask ourselves. What does 'unemployment' look like within the context of a different economy, one in which 'work', reproductive labour, forms of domestic and community care, and the social economy were objects of public policy, recognised and valued? As Simms and Boyle point out there are considerable social and well-being costs to a policy of 'full employment': 'Full employment ... is likely to be corrosive of social capital, if it leaves nobody available in communities' (Boyle and Simms, 2009: 87), and full employment would also likely severely compromise the 'core' or 'social' economy. Full employment, like maximisation and efficiency, are core elements of the project to realise orthodox economic growth. These elements are *not* orientated towards the creation of resilient communities or the enhancement of human well-being. Indeed, as this chapter has pointed out, their collective contribution to orthodox economic growth means that under current conditions (i.e. global and globalising capitalism), they are incompatible with the achievement of these non-economic growth objectives. That is, they are not compatible with a macro-economics of sustainability.

Therefore it seems to me that the real issue here is not the division between 'work' and 'employment' so much as the balance one strikes between them, the pattern of the flow from one to the other, which does require placing both on a more equal footing, rather than seeing 'work' as signifying a lack, or as a less valuable human activity than 'employment', or to be valued only insofar as it supports or leads to formally paid employment. But of course all this leads to the need for a very different conception of the economy than the one that is currently dominant – namely as presented by neoclassical economics. It calls for a much larger, more expansive conceptualisation of the economy in which *all* work, *all* economic activity, *all* resource and energy use is included – not simply those activities which have a monetary valuation or are captured in conventional national economic accounting models. The starkness and ambition of this new green political economy has been captured by Boyle and Simms when they note:

The question is not; as it used to be, how do we make the most profit? It is the broader question of how we create the most human well-being from the least resources whilst living within the thresholds of tolerance of the ecosystems we depend on.

(Boyle and Simms, 2009: 99; emphasis added)

I would add, without exploiting or treating other humans unjustly, compromising democratic freedoms, increasing inequalities, or treating non-humans and the natural world disrespectfully.

Conclusion: beyond growth and beyond carbon-fuelled capitalism

The climate crisis (and associated energy crisis in terms of peak oil) is the defining issue of the current time and is and will continue to shape the politics of the twenty-first century. Taken together with the global economic crisis, the climate and energy crises represent a 'perfect storm' which, if analysed from a crucial political economy point of view in an integrated manner, offer opportunities for radical transformation and transition in thinking and practice in relation to the economy.

In terms of thinking about the 'economy' and 'economics', this chapter has sought to argue that there is a need for critiques of neo-classical economics and its economic 'commonsense' which both depoliticises and naturalises 'capitalism' as the social form under which global humanity and human–nature relations are mediated and constructed. It is for this reason that in thinking about climate change, peak oil, peak water, biodiversity loss and all the other elements of our sustainability crisis, we need to attend to the political economy of capitalism both in theory and practice. Addressing and analysing climate change calls for greater pluralism (and imagination) in our economic thinking, and linking the dominance of neoclassical economic thinking to prevailing asymmetrical power and knowledge relations. These asymmetrical power and knowledge relations should be seen as primary causes of the global sustainability crisis. While I have offered some brief fragments of elements of a green 'macro-economics of sustainability' this clearly needs more development. There are other non/post-capitalist, post-growth political economy perspectives which ought to be explored within a context of greater pluralism, debate and cross-fertilisation within our economic thinking that is so badly needed now.

In terms of the practical politics of a 'macro-economics of sustainability' in relation to climate change, it is clear that there is also some hope in progressive movements and perspectives from the broad Left and Green movements being able to work together around the promise of a low carbon transition being a low- or no-growth one but with high levels of well-being, and above all a shared realisation that an egalitarian agenda is not compatible with orthodox economic growth under capitalism. For example, for the trades union movement this commitment to a 'post-growth' perspective is likely to be too radical and too far removed from its structural commitment to boosting the share of a growing economic pie for workers and their families. However, at the moment where there is common ground and a shared understanding between Left and Green perspectives it is around the reformist 'Green New Deal' agenda and its ecological modernisation promise of combining orthodox economic growth, employment and decarbonising our energy and other infrastructure to deliver a 'step-change' to a low carbon, green economy.

In short, the common green critique of orthodox economics must become a clearer critique of capitalism itself, and relatedly its long-standing and evidence-based critique of economic growth must become a critique of capital accumulation. Any planned economic contraction (in the developed world) as a response to climate change and the transition to a low-carbon energy economy must therefore be viewed for what this is and means: a transition away from capitalism since a non-growth/degrowth capitalism is impossible as well as undesirable. Carbon-fuelled capitalism is destroying the planet's life-support systems and is systematically liquidating them and calling it 'economic growth'. Therefore any plausible macro-economics of sustainability must, at the very least, move us towards a serious examination of radical proposals such as those articulated by eco-socialist (and eco-feminist) movements. A post-growth critique must necessarily lead to a post-capitalist alternative and related political and ideological struggle.

References

Anderson, K. and Bows, A. (2011) Beyond 'dangerous' climate change: emission scenarios for a new world. *Philosophical Transactions of the Royal Society A* 369: 20–44.

Astyk, S. (2008) *Depletion & Abundance: Life on the New Home Front*, Gabriola: New Society Publishers.

Barry, J. (1999) *Rethinking Green Politics: Nature, Virtue and Progress*, London: Sage.

Barry, J. (2005) Ecological modernisation, in Dryzek, J. and Schlosberg, D. (eds) *The Earth Reader*, Oxford: Oxford University Press, pp. 303–321.

Barry, J. (2007) *Environment and Social Theory* (2nd edn), London: Routledge.

Barry, J. (2010) Towards a Green New Deal on the island of Ireland. *Journal of Cross Border Studies* 5: 71–87.

Bowen, A., Fankhauser, S., Stern, N. and Zenghelis, D. (2009) *An Outline of the Case for a 'Green' Stimulus*, London: The Grantham Research Institute on Climate and the Environment.

Boyle, D. and Simms, A. (2009) *The New Economics*, London: Earthscan.

Chaplin, L.N. and John, D.R. (2007) Growing up in a material world: age differences in materialism in children and adolescents. *Journal of Consumer Research* 34: 480–93.

Cutter, S. (2006) *Hazards, Vulnerability and Environmental Justice*, London: Earthscan.

Daly, H. (ed.) (1973) *Toward a Steady State Economy*, San Francisco, CA: W.H. Freeman and Co.

Daly, H. (1993) Postscript: some common misunderstandings and further issues concerning a steady-state economy, in Daly, H. and Townsend, K. (eds), *Valuing the Earth: Economy, Ecology, Ethics*, Boston, MA: MIT Press.

Daly, H. (2003) The illth of nations and the fecklessness of policy: an ecological econo-mist's perspective. *Post-Autistic Economics Review* 22, article 1, available at: www.paecon.net/PAEReview/issue22/Daly22.htm (accessed 1 February 2010).

Degrowth Declaration (2010) available at http://degrowth.eu (accessed 12 December 2010).

Folke, C. (2006) Resilience: The emergence of a perspective for social–ecological systems analyses. *Global Environmental Change* 16 (2): 253–267.

Giddens, A. (1994) *Beyond Left and Right: The Future of Radical Politics*, Stanford, CA: Stanford University Press.

Green New Deal Group (2008) *The Green New Deal*, London: New Economics Foundation.

Hawken, P., Lovins, A. and Lovins, H. (1999) *Natural Capitalism*, Boston, MA: Little, Brown.

Holmgren, D. (2009) *Future Scenarios: How Communities Can Adapt to Peak Oil and Climate Change*, London: Chelsea Green Publishing.

Hopkins, R. (2008) *The Transition Handbook*, Totnes, Devon: Green Books.

Illich, I. (1978) *The Right to Useful Unemployment and Its Professional Enemies*, London: Marion Boyers.

International Labor Organization (2004) *Economic Security for a Better World*, Geneva: ILO.

Jackson, T. (2009) *Prosperity Without Growth*, London: Earthscan.

Kasser, T. (2008) Kasser calls for a revolution in values. *Psychology News*, available at: www.thepsychologist.org.uk/archive/archive_home.cfm/volumeID_21-editionID_157-ArticleID_1302-getfile_getPDF/thepsychologist/0208news.pdf.

Luke, T. (2009) A Green New Deal: why green, how new, and what is the deal? *Critical Policy Studies* 3 (1): 14–28.

McMurtry, J. (1996) *The Cancer Stage of Capitalism*, London: Pluto Press.

Porritt, J. (2005) *Capitalism as if the World Mattered*, London: Earthscan.

Schumpeter, J. (1975/1942) *Capitalism, Socialism and Democracy*, New York: Harper.

Scott Cato, M. (2009) *Green Economics: An Introduction to Theory, Policy and Practice*, London: Earthscan.

Stern, N. (2006) *The Economics of Climate Change*, Oxford: Oxford University Press.

Trainer, T. (2009) Strengthening the vital Transition Towns movement. *Pacific Ecologist* winter: 11–16.

United Nations Environment Program (UNEP) (2001) *Towards a Green Economy: Pathways to Sustainable Development and Poverty Eradication*, available at: www.unep.org/greeneconomy/Portals/88/documents/ger/GER_16_Conclusions.pdf (accessed 12 March 2011).

Wilkinson, R. and Pickett, K. (2009) *The Spirit Level: Why More Equal Societies Almost Always Do Better*, London: Allen Lane.

10 The ideology of growth

Tourism and alienation in Akumal, Mexico

David Manuel-Navarrete

Introduction

Global environmental change and persisting social inequalities pose a challenge to the assumption that economic growth may reduce poverty without creating other negative consequences. A common response to this challenge has consisted of research and discourses on governance reform to allow some sort of reconciliation between economic growth and socio-ecological degradation (e.g. ecological modernization, or inclusive/shared growth). In opposition to such reconciliatory responses, sustainable de-growth acknowledges upfront the impossibility and undesirability of continuous growth. As Barry suggests in Chapter 9 (this volume), the ideology of growth, which is structurally coupled with capitalist political economy, is increasingly identified as a major underlying cause of climate change and natural resources depletion. Therefore, a 'post-capitalist' political economy perspective is needed that questions economic growth.

The least radical de-growth route, often advocated by economists, consists of reforming the dominant economic paradigm. Reform may range from new accounting criteria (that reorient the economy), up to the review of economic paradigms. In either case, the idea is to move away from the goal of 'maximizing growth at any cost' and focus instead in achieving broader welfare objectives, such as full employment, eliminating poverty, and protecting the environment (Victor and Rosenbluth 2007).

A second route to de-growth, perhaps following the work of Polanyi (2001 [1944]), focuses on neither state nor market-based responses. The idea is that both governments and economies would, and ought to, get embedded within empowered local communities which would grant that these institutions work along authentically democratic principles. In this vein, social scientists, including some economists (perhaps regretting some of their professional choices), assert that the transition to de-growth cannot be left to economists or representative governments. Rather, the embedding of the economy into democracy is envisioned as the outcome of political struggle, led by coalitions of social/environmental movements (Martinez-Alier 2009). A comprehensive deployment of this argument is Speth's (2008) call for reinvigorating democracy from below in its deliberative and participatory forms, rather than seeking to direct change from above.

Unfortunately, it is unclear how and why people will disengage from consumerism and take on the burden of democratically designing the rules of the economy. One may argue that there is already an increasing number of people who are deeply concerned about the urgency of climate change and the unfairness of the economy, but this painful realization does not lead them to take any significant action or engage in organized movements (Fournier 2008). Therefore, significant doubts can be cast over the capacity of a de-growth movement to emerge in the short run and wage an ideological struggle against the power of a hegemonic macro-structure, which has emerged from the interplay between market economy, capitalist production conditions, and representative democracy. This macro-structure is tremendously sturdy and confers stability to the tyranny of growth both ideologically and practically, while at the same time depending on this very tyranny for its reproduction.

A third route proposes 'inclusive democracy' as a universalist project for human liberation and autonomy. Drawing on libertarian thought, this route postulates a fundamental incompatibility between liberal market economies and de-growth (Fotopoulos 2007). Accordingly, strategies based on changing behaviour, democratic reform, or radical decentralization are not seen as effective, or even realistic, because they do not address either power relations, or the historical characteristics of capitalism, which are both structurally linked to the need for continuous growth. Therefore, a pre-condition for de-growth is a more egalitarian distribution of power, which would result from a confederation of communities functioning according to principles of economic equality, collective ownership, and direct democracy.

Again, the means, or praxis, to achieve this organizational vision is the crucial issue. Inclusive democracy does point to a conscious and self-reflective choice for autonomy (over heteronomy) as a driving force towards de-growth. Perhaps following Hegel, Fotopoulos (1997: 181) situates this choice in the dialectical tension between individual versus collective, agency versus structure, or idealism versus materialism. The origin of this tension is situated in the fact that individuals are both free to create their world and at the same time are created by the world. According to Fotopoulos (2000), the way out of this tension is a liberatory project towards individual and collective autonomy that synthesizes democratic, socialist, libertarian, green and feminist traditions.

In this chapter I discuss the usefulness of the concept of alienation in the study of growth. This exploration is based in bibliographic and ethnographic research that sought to reconstruct life-story narratives (see Bruner 2004) of agents promoting tourist growth. The development of two adjacent coastal enclaves in the Mexican Caribbean, a region subjected to intense tourist commoditization, is compared to illustrate the usefulness and difficulties of researching alienation.

Alienation and the ideology of growth

Alienation is largely absent from de-growth debates. However, it has been tangentially addressed through notions such as human development, self-emancipation and autonomy from capital (Lipietz 1995; Holloway 2002; Burkett 2005). In his seminal piece about the meanings of alienation, Seeman (1959) identified five ways in which the concept has been used in social science, including alienation as: powerlessness, meaninglessness, normlessness, isolation, and self-estrangement. Powerlessness, mainly understood as the outcome of the separation of workers from the means of production, has probably persisted as the most common usage in socio-political literature. However, I will draw on the fifth interpretation: alienation as self-alienation, which Seeman associates with the work of Erich Fromm, C. Wright Mills, Nathan Glazer and others. In particular, I will build on Fromm's (1961) emphasis on the process of commoditization (or objectification) of the entire world as the root of alienation, and will complement it with Schmitt's (2004) definition of alienation as the inability of individuals to forge an intelligible history about their own lives.

My interest in Fromm's viewpoint lies in that (perhaps inspired by Max Weber) it goes beyond the emphasis on industrial production to include a broader scope of human experiences. Thus, humans are not mainly situated within the productive process, but within the world at large. As a consequence, alienation is broadened to take into account separation from ourselves (i.e. from our power as self-creating individuals), society and nature. In this sense, alienation has its origin in the self–society–nature interaction as it results from individual selves' powerlessness to influence the constraints and demands imposed upon them by nature and social structures. In his analysis of Marx's Economic and Philosophic Manuscripts (1964), Fromm discusses alienation in the following terms (Fromm 1961: 42–43):

> It is not only that the world of things [commodities] becomes the ruler of man, but also that the *social* and *political circumstances* [society] which he creates become his masters. … The alienated man, who believes that he has become the master of nature, has become the slave of things and of circumstances, the powerless appendage of a world which is at the same time the frozen expression of his own powers. For Marx, alienation in the process of work, from the product of work and from circumstances is inseparably connected with alienation from himself, from one's fellow man and from nature. … He is alienated from the essence of humanity, from his 'species-being', both in his natural and spiritual qualities.

One limitation of Fromm's approach is its high level of abstraction. In order to make it more empirically researchable, I will combine Fromm's with Schmitt's (2004) notion that alienation takes place when individuals fail to find meaning in their own lives. We seek meaning within groups we belong to, established traditions or family values, but when we blindly follow external rules and conventions without reflecting on them we stop being ourselves and became alienated from our essence. According to Schmitt (2004), to be oneself means, first, managing the different personalities that one can potentially develop, while at the same time presenting a reasonably clear identity to others. Second, it means having a certain degree of ownership over the direction that one gives to the development of one's own personality.

The combination of the two approaches results in a notion of alienation that considers the personal power to give direction to one's own life, to build one's own life-story narrative, but also the need to critically assess how these narratives, and the social structures enabled by them, constrain ourselves and others to function at a distance from nature, community, and our/their own creations and selves (Kalekin-Fishman 2006). This positioning may raise objections on the basis that every person has the right to give sense to what matters in their own lives and that this personal decision should not be subjected to criticism. Obviously, an approach that seeks to censor the election of personal narratives would only contribute to further alienation. What I am proposing here is to open life-story narratives to reconstruction, commentary and critique based on the fact that the planet is finite, that humanity cannot materially grow forever, and that the accumulation of capital by a few shrinks the possibilities of the majority to shape their own narrative because it dramatically limits, among other things, their choices for material subsistence or fulfilment.

The following sections seek to illustrate, empirically, the connection between alienation and tourist growth in two adjacent tourist enclaves in the Mexican Caribbean: Akumal Playa and Bahia Principe.

Urban development did not start in the northern part of the Mexican Caribbean until 1974, when Cancun emerged as the first integrally planned centre for mass tourism in Mexico (Martí 1985). From the 1980s onwards, demographic and economic growth has been intense and closely linked to tourism. Official censuses indicate a tenfold population increase between 1970 and 2000 while international tourist arrivals doubled from 3,070,695 in 2000 to 6,113,705 in 2008 (SECTUR n.d.). Akumal is located about 100 kilometres south of Cancun, between the booming towns of Playa del Carmen and Tulum, in a stretch of coast commercially known as the Maya Riviera. Akumal Playa was the first tourist enclave to emerge in the Maya Riviera. Today it hosts about 1,200 residents and is flanked by all-inclusive resorts owned by Spanish corporations. The case of Akumal is adequate to study the relationship between alienation (from self, society and nature) and tourist growth because its development was tightly led by a single individual, Don Pablo Bush and his family. Arguably, this circumstance facilitates a parallel discussion between social relations fostering or inhibiting tourist growth, and the subjective trajectories related with alienation of key individuals. Likewise, the Bahía Príncipe is the largest all-inclusive resort of Quintana Roo, but it was developed by a corporation self-defined as a family business and tightly led by Pablo Piñero. Discussion in the following sections will seek to reconstruct the life-stories of the main leaders of tourist growth in these two enclaves in order to speculate about the relationship between alienation and the hegemony of growth.

Akumal Playa

Akumal was created in the 1960s, more than a decade before tourist mega-developments sprouted in Cancun (Figure 10.1). Currently, the enclave is divided into Akumal Playa: a resort containing about a dozen small hotels and 100 high-end houses, mostly owned by US citizens; and Akumal Pueblo: located directly across the road and populated by immigrant workers who came largely from the neighbouring states of Yucatan and, more recently, Chiapas. Akumal's 'founder', the late Don Pablo Bush, was a Mexican entrepreneur, explorer and adventurer, who in 1962 acquired concessionaire rights over a coastal property which was once a copra plantation. This piece of relatively pristine Caribbean coast had already been 'discovered' by a few US diving expeditions in search of wrecked ships and sunken treasures. Concerned with the risks of looting, Don Pablo led the creation of CEDAM (Exploration and Aquatic Sports Club of Mexico) to rescue the *Matancero's* shipwreck remains. At the same time Don Pablo created a small hotel and a *palapa* restaurant to host the few visitors attracted by marine life exploration and the beauty and tranquillity of the coast.

Before Don Pablo's arrival, the area had already plunged into global commodity markets and the extractive logics of capital through the exploitation of copra and marine resources. However, these activities were in decline in the early 1970s when Don Pablo founded the Promotora Akumal Caribe, a corporation to divide and sell the land for residential development (Periódico Oficial del Estado, 20 January 1975). At around the same time he founded the Akumal Yacht Club with friends and entrepreneurs from Mexico City and other parts of

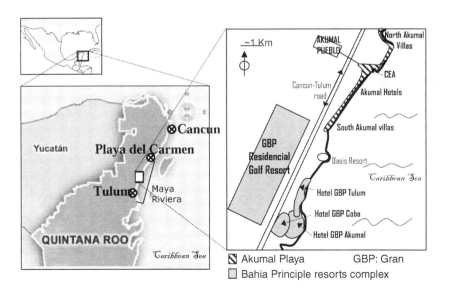

Figure 10.1 Location of Akumal and Bahía Príncipe resorts in Quintana Roo, Mexico.

North America. The club gradually inceased to 33 associates and congregated a group of tourist entrepreneurs who would later share a 'green version' of Quintana Roo's industrial tourist model, a vision that has been described in terms of ecological modernization (Redclift *et al.* 2011). Today, they are powerful local businessmen, including Román Rivera Torres (developer of Puerto Aventuras), Francisco Córdova Lira and the Constandse brothers (developers of the all-inclusive eco-parks X-Caret and Xel-Há, among other projects). Already in the 1970s, this group operated a construction company (RITCO Group), which was instrumental in the development of Akumal and grew fast thanks to the rise of Cancun.

From the point of view of alienation from nature, Akumal's development may be characterized by the dialectics between two impulses. On the one hand, the awe that Don Pablo and his entrepreneurial associates felt for this wondrous landscape and wildlife. In this sense, their decision to settle in Akumal may be seen as a radical way of reconnecting with nature by leaving their urban life-styles behind. On the other hand, such awe-inspiring scenery may have also indicated the potential to attract wealthy people and thus guarantee capital flows, financial growth and material security. The main motivation as alleged by an interviewed club member was 'improving our quality of life and searching for adventure'. In fact, Akumal represented for some of them a collective and auda-cious 'life's project' whose success depended on a delicate balance between nature conservation and selective engagement with capitalist endeavours. This project's success would ultimately depend on the challenging task of keeping the area away from the industrialist model of tourist growth, which is prevalent in Quintana Roo, while enabling low-density, high-end development.

The sense of adventure and ownership of Akumal's development felt by Don Pablo and, to a large extent, by the other club members may be interpreted as providing a sense of ownership over the making of history. However, this project was necessarily constrained by the global and national socio-political structures that have conditioned the growth of tourism across Quintana Roo. It was also obviously influenced by the personal history (for example, as professionals, busi-nesspeople and owners of capital) that club members brought with them to Akumal. In addition, ownership of land, beyond one's own sheltering needs, involves the alienation of others: by disallowing them to experience certain aspects of nature or significantly restricting the scope of their own 'life's projects'.

The discussion above shows how autonomy, as perceived and enacted by some, is often connected to the alienation of others. In Quintana Roo, tourist growth is often coupled with the (officially illegal) exclusion of local people from beach access (Manuel-Navarrete and Redclift 2011). In addition, uneven development, asymmetric accumulation of capital and unstable job markets produce social inequalities which shape daily life and social interactions (Torres and Momsen 2005). In the beginnings of Akumal as a tourist resort, business owners, tourists and workers would intermingle personally and interact closely. However, in 1992 club members decided, perhaps mimicking Cancun's model, to create a new town 5 kilometres away, today's Chemuyil, and relocate workers there. This displacement was justified, and still is today, by both government agencies and local businesses as a need to protect the commercial image of the resort. However, it produced a spatial differentiation which alienated workers from their original community and access to the beach (now designated as space for tourist use), while forcing them to commit to a mortgage that in most cases would take the work of a lifetime to repay. Those workers who did not want to move to the new town managed to resettle themselves a few hundred metres from Akumal Playa, on a piece of federal land, today's Akumal Pueblo.

There is little doubt that Akumal's landscape has been significantly altered by tourism. However, the club's gatekeeping role prevented the area from being taken over by all-inclusive impersonal resorts. By all means, Don Pablo had a strong ethical, at times paternalistic commitment to protect the very image of Akumal that he had himself created; that is, the image of an exclusive and environmentally preserved place. His daughter illustrates through an anecdote how at times he used to feel disappointed:

> He thought that Akumal would be a small place, semi-private, without mass tourism or [snorkelling and diving] tours in the bay. Now there are many boats. He never imagined that there would be three-storey condos. Once, when he was already old, I took him to Yal-Ku lagoon and when he saw a three-storey condo, he said: Why is that there? Why has it not been pulled down? For him it was illogical that someone could build such a building in a land that he had classified as low density.

This sentiment towards Akumal is consistently shared, at least to some extent, with other members of the club. This highlights the importance of place attachment for strategies to resist the hegemony of economic growth. Palmira, again, illustrates this point quite graphically:

> the mentality for conservation came to me here in Akumal, the moment I saw all the beauty. I was not looking for it because I was hardly starting to get tired of Mexico City. However, when I started to live in one of the bungalows by the beach I thought: I would not change this for anything else, and I had a luxurious penthouse downtown in Mexico City. The thing is that nature talks to you ... we had decided to come to live here, we were raising our kids here. ... Laura and Oscar were fundamental because they were the ones owning local companies and therefore they were the kind of people who would say: I am from here! One feels the ownership over the land, I belong here! Although I have always said that I am not owner but guardian of the place where I live.

In 1993 the Yacht Club was reconverted into an ecological centre. Thereby the *Centro Ecológico Akumal* (CEA) was created including a generous donation of prime land (4 ha) and buildings from the club. The rental of the buildings and space owned by CEA provides a steady source of income for the centre's programmes and activities which allows some continuity for conservation programmes and actions and enables long-term planning. The ability to formulate and implement local environmental strategic planning may increase Akumal's climate adaptation capacity through, for instance, the maintenance of key ecosystem services that offer protection from hurricanes and storms (Manuel-Navarrete *et al.* 2010). In this sense, as analysed in Manuel-Navarrete *et al.* (2010), the development vision underlying tourist growth in Akumal may be seen as counteracting the hegemonic developmentalist vision which prevails across the Mexican Caribbean, and as providing a higher diversification of adaptation and coping strategies.

Bahía Príncipe hotels and resorts

To the south of Akumal, Bahía Príncipe's three hotels and Residential Golf Resort constitute the largest all-inclusive enclave in Quintana Roo (Figure 10.1). It features approximately 3,000 rooms, hundreds of residences (each ranging from 250 to 650 square metres) and a 90 ha golf-course. Bahía Príncipe Clubs & Resorts is the youngest division of the Piñero Group, a family-owned transnational tourist corporation founded by Pablo Piñero in 1975 and based in the Balearic Islands (Spain). Like most Balearic transnational tourist corporations, the Piñero Group has grown through a strategy of vertical integration. Thus, the group started with a wholesale tour operator (Soltour) which then integrated the management of hotels, building companies and other tourist services in order to gain a foothold on almost every step of the tourist chain. This strategy reduces risk and vulnerability by ensuring that none of the chain's links will take advantage of market contingencies, as happened in the Balearic Islands during the oil crises of the 1970s when reduced profit margins exposed hotel managers to wholesalers taking advantage from their control over pricing policies (Bray and Raitz 2001).

The transnationalization of Grupo Piñero started in the 1990s with the replication, in the Dominican Republic and Mexico, of the all-inclusive model pioneered in these same countries by fellow Balearic corporations such as Sol-Melia and Barcelo. This model consists of holiday towns offering multiple tourist services and enabling parallel real estate developments, the so-called condo-hotels. The goal is to build large-scale hotel infrastructure so that the tourist is confined within the resort. Governments, eager to attracting international investment, often provide benefits and incentives to promote this type of development, including tax cuts, liberalization of foreign currency exchanges, basic infrastructure provision, low land prices, and the ability to settle contracts in foreign currency (Buades 2009).

Pablo Piñero likes to portray himself as a 'self-made' businessman. A former public sector employee, his 'adventure' into the tourist business started when he organized a charter flight from his native Murcia (in continental Spain) to

Majorca in 1970. Bahía Príncipe hotels in Akumal started to be built in 1998, while the huge Riviera Maya Residential Golf Resort began construction in 2006. This 'adventure', however, seems of a different quality to Bush's in Akumal, and, independently from Piñero's original motivations, the outcome of 40 years of unrelenting growth is a sizeable economic empire which employs about 10,000 people and is worth about €2,000 million, including branches in Portugal, Russia, the Caribbean and the USA (Nexotur 2011).

Grupo Piñero is not a large tourist corporation by international standards. Among Spanish corporations, it ranks only fifteenth in number of managed hotel rooms (Such Devesa 2004). However, it ranks tenth in the percentage of rooms operated abroad, and its internationalization since 1995 has been very rapid. In 1998 Bahía Príncipe Clubs & Resorts started its expansion in Mexico with an initial investment of about US$100 million. Between 1998 and 2005 the group acquired, from the government of Quintana Roo, 477 ha of land to build three hotels and residential resorts for an average price of US$2.24 per square metre (Carrera 2006). As Pablo Piñero recognizes, this was an extremely favourable bargain.

Arguably, unlike Don Pablo Bush, Piñero was not considering his investments in the Mexican Caribbean as a platform for a new 'life's project'. Beyond his likely appreciation of the beauty of the Maya Riviera's landscape, his motivations were rather framed within the logics and language of a business project towards corporate growth. He claims that the key for the meteoric growth of his company is a combination of good luck and hard work. He belongs to a new bourgeois class of Balearic families who took advantage of the boom in Fordist mass tourism that emerged in Europe during the 1960s (Bray and Raitz 2001). Their national consolidation was supported by Franco's dictatorial regime due to the role of tourism in adjusting the country's foreign trade imbalances in the 1960s (Buades 2006). Once nationally consolidated, these emergent family corporations benefited from neoliberal deregulation to become inserted within global financial markets and expand internationally through hotel, real estate and tour-operator holdings. Corporative alliances and novel global financial products such as real-estate investment trusts or hedge funds significantly increased the possibilities for expansion of these family businesses.

It is important to note that this Balearic new bourgeoisie does not emerge from an aristocratic or industrial class, but from the bottom. Their success is partially due to their vocation for work, willingness to seize business opportunities and ability to establish convenient relations with governments. Like some sectors of the American industrial bourgeoisie, they seem to share an epic feeling characteristic of pioneering endeavours. The objective conditions for the emergence of this new class are to be found in the Fordist industrialization of Europe which allowed for an overwhelming participation in tourism, fuelled by cheap flights, of the middle and working classes, which would later become an elite of global travellers. However, the internationalization of the class is a post-Fordist phenomenon fuelled by low wages, financial innovation, and a global socio-political context enabling growth at any cost and as fast as possible.

Pablo Piñero is eager to highlight his 'vocation of work' (López 2009: 41): 'I do not know how to do anything else, I do not have another hobby but to work and I will die with my boots on. I will be working here while my health allows me.' When asked about his dreams, he shows little concern for current debates on retirement age; he wishes 'to arrive at 85 years old here in the office' (López 2009: 41). The 'vocation of work' of this allegedly 'self-made' class can be related to 'Americanism', as described by Gramsci in the early twentieth century. According to Gramsci, Americanism was manifested in the fact that a millionaire continued to be practically active until forced to retire by age or illness and that his activity occupied a very considerable part of his day. He speculated about the origin of Americanism in terms of:

> [A] recent 'tradition' of the pioneers, the tradition of strong individual personalities in whom the vocation of work had reached its greatest intensity and strength, men who entered directly, not by means of some army of servants and slaves, into energetic contact with the forces of nature in order to dominate them and exploit them victoriously.
>
> (Gramsci and Forgacs 2000: 293)

The Balearic tourist bourgeoisie did not inherit, at least not directly, the pioneer tradition referred to by Gramsci, but they have probably been inspired by it through direct exposure to American corporate culture. Arguably, they were able to experience, first hand, the pioneering spirit through their expansion to the Americas, an expansion that many local people refer to as '*reconquista*' (reconquest). At the least, the USA is perceived as an enabler of business internationalization. In Piñero´s words (López 2009: 41):

> The best area for business, without any doubt is the Caribbean. It is very close to the USA, and for that reason it is its backyard and it is guaranteed that there will not be any trouble. ... Prices are lower everyday due to aviation and there is no Islamism.

Conditions of workers in Bahia Principe hotels and resorts are highly precarious. Most workers live in Tulum or Playa del Carmen, but some need to travel for over two hours to get to work. The contrast between luxurious tourist grounds and workers´ residences is stark. However, there is little doubt that these workers are no less alienated from the product of their work than in any other capitalist company. Perhaps more significant, though, is that in exchange they obtain wages which barely allow subsistence and payments towards a mortgage to buy a basic house on the outskirts of Tulum or Playa del Carmen.

The construction of Bahía Príncipe prompted the creation of an environmental group, SAVE Riviera Maya, led by US citizens inhabiting or owning houses near the macro-resort development. This group insistently denounces the destruction of mangroves and the geo-engineering practices in the beach adjacent to the hotels that seem to be conducted by the resort with the purpose of enlarging the areas of sand, 'improving' the aesthetics of the landscape, or 'combating' erosion from hurricanes and storms. Perhaps in response to SAVE's attacks, or maybe as part of the general trend to improve corporate social responsibility among Balearic tourist corporations, the Ecological Foundation Bahía Príncipe Tulum was created. The foundation started a programme in 2000 for the study of the coral reef and conservation of turtles, and an environmental education programme in 2005. It is clear, though, that the role of the foundation is not comparable to that of CEA in Akumal. Despite the benefits derived from its programmes, the foundation is not a significant player in the governance of Bahía Príncipe.

Conclusion

The life-stories constructed around tourist growth in Akumal Playa speak about the search for meaning in the experiencing of a place, which is generally valued as a means for urbanites to reconnect with nature, or with an idealized idea of nature. The Bush family and their associates experienced this connection through specific activities, such as diving, but also through the idea of inhabiting and owning the place. At the same time the formal ownership of the land provided the means to avoid alienation from global markets, capital and all the benefits of modern societies. Thus, Bush's may be judged as a narrative situated in the tension between these two forms of alienation (participation in nature versus participation in capital) and the search for meaningful ways of balancing them. The Yacht Club allowed others to join in the unfolding of the narrative, as well as consolidating and institutionalizing it. Maya immigrant workers, however, were alienated. Social alienation within Akumal was necessary for the exploitation of labour. This exploitation was in turn instrumental to enable Bush's and others' balance between certain forms of connection to nature and highly profitable forms of connection to capital. Social alienation was institutionalized with the actual separation between tourist and residential space. However, Akumal Playa is still today one of the few public beaches of the Mayan Riviera and one of the most used by locals. In opposition, all-inclusive resorts threw up numerous physical obstacles for public access and are actually almost exclusively used by hotel guests.

The life-story narrative constructed around Piñero's pseudo-biography does not draw meaning from his personal involvement in Bahéa Príncipe as a place or in any close relationship with his 2,000 employees. Rather, the language of growth ideology seems to dominate a narrative about the objectification of Bahía Príncipe in terms of, for instance, investments and returns, the representation of workers as numbers and the role of nature as a factor of production. Piñero's personal narrative is one of hard work, success and prestige measured in terms of the economic growth achieved through his group. An interesting question is to what extent this narrative may be considered as originally created by Piñero, or on the contrary is a performance narrative borrowed from others or from a global script. In any case, Piñero's is an intelligible narrative shared and sought by many. In fact, a key element in its unfolding is the participation of hundreds of thousands of middle-class tourists from Europe and North America. A crucial question for alienation research on tourism is whether corporations create all-inclusive resorts to respond to a fashion imposed by 'the market', or whether they are actually the main driving force creating such fashions. By the same token, another crucial inquiry is the autonomy of corporations to set the labour conditions of their employees. Is the constant drive to keep wages as low as possible compelled by impersonal and inescapable economic laws or by the employees' alienation?

From a standpoint of self, neither Bush nor Piñero are cases of self-alienated individuals in the sense of having little autonomy to influence the course of their life-stories, or of lacking relative control over the external conditions shaping them. However, as discussed above, significant differences emergy when comparing how each actor incorporates and represents the social and natural contexts where tourist growth takes place. These differences may have significant implications for the analysis of global environmental change and social inequality in the context of the multiples crises of capitalism. We have explored the different qualities of the connection with local ecosystems for each case. In the case of Akumal Playa a closer personal connection has resulted in the prominence of an environmental actor, the Akumal Ecological Centre (CEA), which plays a key role within local governance arrangements. We have speculated on whether this key role increases local capacity to climate adaptation such as coping with hurricanes and storms. However, despite their differential local environmental impact, both Akumal Playa and Bahía Príncipe are not only dependent on constant flows of global capital and plentiful availability of cheap labour; they are also highly carbon intensive through the importing of goods and bringing in of tourists. These global dependencies are possibly acknowledged by Bush and Piñero, but at the same time taken for granted and therefore not reflected upon as significant variables in their life trajectories. They tend to be regarded as contextual (alien?) factors over which little control can be exercised. Yet, the contemporary combined crisis of global environmental change and the global economic downturn are precisely calling our attention to the need to reflect further upon these factors. The question remains of how these global interdependencies can be meaningfully integrated into our life trajectories instead of alienating.

References

Bray, R. and Raitz, V. (2001) *Flight to the Sun: The Story of the Holiday Revolution*, New York: Continuum.

Bruner, J. (2004) Life as narrative. *Social Research: An International Quarterly* 71 (3): 691–710.

Buades, J. (2006) *Exportando paraísos. La colonización turística del planeta*, Palma de Mallorca: La Lucerna.

Buades, J. (2009) *Do not Disturb Barceló*, Barcelona: Icaria.

Burkett, P. (2005) Marx's vision of sustainable human development. *Monthly Review* 57 (5): 34–62.

Carrera, V. (2006) Bahía Príncipe, de Pablo Piñero, el mayor beneficiado del remate de predios del Fidecaribe e IPAE. *Noticaribe*, 12 November.

Fotopoulos, T. (1997) *Towards an Inclusive Democracy: The Crisis of the Growth Economy and the Need for a New Liberatory Project*, London: Cassell.

Fotopoulos, T. (2000) Beyond Marx and Proudhon. *The International Journal of Inclusive Democracy* 6 (1). Available at http://www.inclusivedemocracy.org/journal/.

Fotopoulos, T. (2007) Is degrowth compatible with a market economy? *The International Journal of Inclusive Democracy* 3 (1).

Fournier, V. (2008) Escaping from the economy: the politics of degrowth. *International Journal of Sociology and Social Policy*, 28 (11/12): 528–545.

Fromm, E. (1961) *Marx's Concept of Man*, New York: Frederick Ungar.

Gramsci, A. and Forgacs, D. (2000) *A Gramsci Reader: Selected Writings, 1916–1935*, London: Lawrence & Wishart.

Holloway, J. (2002) *Change the World Without Taking Power: The Meaning of Revolution Today*, London: Pluto Press.

Kalekin-Fishman, D. (2006) Studying alienation: TOWARD a better society? *Kybernetes* 35 (3/4): 522–530.

Lipietz, A. (1995) *Green Hopes*, Cambridge: Polity Press.

López, D. (2009) Pablo Piñero Imbernón: El mayor error es no hacer 8 o 10 campos más de golf en baleares. *Masmagazine* 54: 36–41.

Manuel-Navarrete, D. and Redclift, M. (2011) Spaces of consumerism and the consumption of space: TOURISM and social exclusion in the 'Maya Riviera', in Pertierra, A. and Sinclair, J. (eds) *Consumer Culture in Latin America*, Basingstoke: Palgrave Macmillan.

Manuel-Navarrete, D., Pelling, M. and Redclift, M. (2010) Critical adaptation to hurricanes in the Mexican Caribbean: development visions, governance structures, and coping strategies. *Global Environmental Change* 21: 249–258.

Martí, F. (1985) *Cancún, fantasía de banqueros: La construcción de una cuidad turística a partir de cero*, Mexico City: Unomasuno Editorial.

Martinez-Alier, J. (2009) Socially sustainable economic de-growth. *Development and Change* 40 (6): 1099–1119.

Marx, K. (1964) *The Economic and Philosophical Manuscripts*, New York: International Publishers.

Nexotur (2011) 13 January. El presidente del grupo Piñero, Pablo Piñero, nombrado prócer del turismo español en Iberoamérica en CIMET. *Nexotur.com*.

Polanyi, K. (2001 [1944]) *The Great Transformation: The Political and Social Origins of Our Time* (2nd edn), Boston, MA: Beacon Press.

Redclift, M., Manuel-Navarrete, D. and Pelling, M. (2011) *Climate Change and Human Security: The Challenge to Local Governance Under Rapid Coastal Urbanization*, Cheltenham: Edward Elgar.

SECTUR (n.d.). Datatur. Retrieved May 2010 from: http://datatur.sectur.gob.mx: Secretaría de Turismo de México.

Seeman, M. (1959) On the meaning of alienation. *American Sociological Review* 24 (6): 783–791.

Schmitt, R. (2004) *Alineación y libertad*, Quito: Ediciones Abya-Yala.

Speth, J.G. (2008) *The Bridge at the Edge of the World: Capitalism, the Environment and Crossing from Crisis to Sustainability*, New Haven, CT: Yale University Press.

Such Devesa, M.J. (2004) *La financiación del sector hostelero español: Aspectos financieros de la expansión internacional de las cadenas hoteleras españolas*, Alicante: Biblioteca Virtual Miguel de Cervantes.

Torres, R.M. and Momsen, J.D. (2005) Gringolandia: THE construction of a new tourist space in Mexico. *Annals of the Association of American Geographers* 95 (2): 314–335.

Victor, P.A. and Rosenbluth, G. (2007) Managing without growth. *Ecological Economics* 61: 492–504.

Part IV

The new politics of climate change

11 Utopian thought as a missed opportunity and leverage point for systemic change

Mattias Hjerpe and Björn-Ola Linnér

Introduction

Why are the visions of development that incorporate climate change so constrained? It may be argued that conservative futures thinking is less likely to alienate and more likely to offer easy connections to current policy and practices. This may be true in part, but to date the failure of national actors and international fora to squarely address climate change as a development issue suggests we need to look for alternative framings. The likelihood of failure is clear, for example, in the work of the Climate Action Tracker which calculates that even if we implement the most stringent reductions proposed at the UN Climate Conference in Cancun in 2010, we would still miss the 2°C target by 8 Gigatonnes (Gt) CO2-equivalents annually by 2020. This conclusion is supported by similar gaps identified in other reports (UNEP 2010). Utopian thinking offers an alternative logic, one that sets its sights high, maybe unachievably high, but in so doing forces a consideration of a wider range of choices that need to be made to confront the root causes of climate change.

There is no doubt that achieving the 'deep cuts in global emissions' (UNFCCC 2009) calls for systemic changes of production and consumption patterns, which requires leapfrogging in innovation and/or lifestyle changes. The meagre track record of the current cap-and-trade path under the UN Framework Convention on Climate Change (UNFCCC) may at best be characterised as a tiny but costly adjustment of the carbon society. Accordingly, an increasing number of analysts are concluding that this is a dead-end. The Hartwell paper on a new direction for climate policy called the UNFCCC model 'structurally flawed and doomed to fail' (Prins *et al.* 2010: 5).

How then can a systemic change be achieved? In this chapter we turn to utopian thinking as a leverage for envisioning new alternatives that could unleash societal creativity in futures thinking in relation to climate change. We contend that utopian thinking has been underestimated as a tool for understanding self-organised limitations when thinking about sustainable and decarbonised futures that go beyond business as usual and enable a discussion on systemic change. We apply Kumar's distinction between utopian proper and utopian thought to analyse if and how elements of utopian logics have surfaced in climate change science and policy. This chapter presents examples that could potentially serve as a springboard for utopian thinking in climate change politics.

Our findings are drawn from a reading of decisions and documents retrieved through the official UNFCCC documentation service, and abstracts from official side events at the UN Climate Change Conferences, some of which have been presented in Hjerpe and Linnér (2009). This material captures the main policy debates about futures and debates within research, environmental NGOs, business and indigenous people's organisations. Based on this, we have selected three examples that we argue could each potentially serve as a springboard for utopian thinking in climate change politics and one example that illustrates how alternative futures are masked in the promotion of particular technical options. Complementary material, in particular reports elaborating the programmes, was collected from organisations' websites.

Utopian thought

A utopian momentum

Why focus on utopian thought? Utopian thinking has amassed negative connotations as a result of totalitarian efforts to bring about utopias in real polities accompanied by claims of absolute truth (Hedrén and Linnér 2009). In the wake of Nazi efforts to construct a racist utopia and Stalinist oppression in the name of the dictatorship of the proletariat, utopia has become an emblem of the dangers of ideological conviction for many liberal analysts. Scholars have also explained the lack of utopian expression in contemporary debates as a feature of post-modernism's dislike of absolutes made material by the multiple tensions inherent in global capitalism (Kumar 2003). We argue that to reclaim its relevance and potential as a conceptual driver for transformation, utopia needs to transcend three fundamental and problematic aspects of modernity: scientification, or the notion of fixed truth; nationalism, or the notion of fixed space; and blueprints, or the notion of fixed political goals (Hedrén and Linnér 2009). It was the failure to overcome these elements of utopia during the twentieth century that led to its capture by dogmatic and oppressive political orthodoxies.

The challenge is to see how far it is reasonable and possible to move beyond the inherited political concerns around utopian thinking and from here to expand the imagined possibilities of what society could be, in this way justifying an escape from the technical realism framing that has so far dominated futures thinking on living with climate change – and concatenated economic or political risk and uncertainty (Jameson 2004; Linnér and Hedrén 2009). The potential for utopian thinking to work as a springboard for social change has already been noted and, as Giddens (2009: 12) reminds us, Reverend Martin Luther King did not spur activity by calling out 'I have a nightmare!' Surfacing contrasting aspirations and dreams about the future could potentially vitalise the environment-development debate and open new opportunities to think about and live alternative socio-ecological futures (el-Ojeili and Hayden 2006).

Utopia proper and utopian thought

The word *utopia* is derived from the Greek words for 'no place' and 'good place'; the ideal good place is also no place. Kumar distinguishes general utopian thought from utopia proper: the former distils the essence of a vision but remains conceptual; the latter provides a narrative of a hypothetical society including its functional features (Kumar 1999). Following Kumar, this chapter uses utopian thought and utopian proper to explore the existing elements of utopia in climate politics and to point towards openings for a greater role. Utopian thought and utopian proper can be differentiated by the degree to which they express specific codes of morality and preferred pathways for transformation.

Morality

All utopias present a vision of a good society underpinned by a clear moralistic message. To be genuinely utopian, visions require an explicit morality; a clear message of what social action is good and bad, right and wrong (Hedrén and Linnér 2009). In this way utopian thought and utipian proper contain detailed prescriptions for society (Kumar 2003; Hedrén and Linnér 2009). Utopian thinking outlines the logics for rationally and secularly: how the good society should be designed (Kumar 2003). It describes a fictitious, ideal place where society is organised to handle current challenges and contradictions. Here the role of utopian thinking is to paint pictures of daily life, instead of abstract formulations (Kumar 2003).

Western utopian thought may be grouped into three traditions on the basis of moral content and value prescriptions: social egalitarianism, technological optimism and natural or indigenous romanticism. These three variants sit at the boundaries of Western society's technical, moral and social imaginations. Social egalitarianism has been evoked to legitimise recent, historic socio-technological transitions. Examples include dreams of a leviathan-style global government made possible by the interdependence of the global industrial and post-industrial economy, or aspirations for an ultra-modern and efficient society led by technology. Indeed, technological development has often appeared central to Western utopian thought, providing a catalyst for thinking about radically different social futures. In technological utopias it is the employment of advanced science and technology that transforms society; material living standards are raised for all with everyday activities such as sleeping, eating and travelling radically affected at the heart of a utopian vision, though at times accompanied by a dark side of reduced personal freedom. This is a fear illustrated, for example, in the dystopic vision of Huxley's *Brave New World*. Perhaps in response to technological utopia, the 'Arcadian' (Kumar 1987) or indigenous utopian visions (Jorgensen 2005) have been presented that replace technology as a driver for change and basis of legitimacy by traditional, 'authentic' practices and material evolved over millennia in apparent harmony with nature. Indigenous versions of utopia may be conservative and romantic, recalling the lost, imagined pasts like that of the

Garden of Eden. A connected strand of utopian imagination draws from ecological imagery. Here the relation between people and nature is central. People though are seldom central, for example, in visions inspired by deep green philosophy where humanity is considered one among many species with equal rights claims to life and security. Such ecological utopias are especially popular in political discourse at times of perceived resource scarcity.

In the twentieth century, critical utopian thought has been most prevalent amongst the neo-Marxist tradition. Importantly, this school has drawn upon ideas of an egalitarian utopia offered by the sixteenth-century political theorist Thomas More (Wallerstein 1986). Neo-Marxist utopian thinking argues that the logic of capitalism promotes *a*topia, i.e. the absence of place instead of no place. This is achieved through the free movement of capital, production factors, goods and services. Alienation of people from places where they may live, from the places where goods they consume originate or are manufactured, or where wastes accumulate is argued to express itself in feelings of helplessness and the inability to change the status quo (Hardt and Negri 2000). In short, the alienation of self.

Pathways to transformation

More's framing has utopia as a non-existing place, located at the very edge of our understanding (Jameson 2004) – an experiment in imagination with political implications but only indirectly material consequences. Others have been more ambitious. Herein is a challenge for utopia, as noted above. In line with Hedrén and Linnér (2009), we argue that the experience of twentieth-century Western utopia proper is a clear warning that utopia should not be seen as a policy blueprint for action. Rather utopia acts best as a heuristic means for inspiration. The boundary between philosophy and practice is not always clear, with Kumar (2003: 2) noting that past utopian thinking has 'displayed a certain sobriety' but that 'it is never simply dreaming. It always has one foot in reality'. The strength of utopian thought is the ability to think beyond the material constraints and cultural habits of the present and in this way to influence society (cf. Block 1918).

In distancing itself from the present, utopian thinking can also present pathways for transformation (Wallerstein 1986) – in technology, social organisation or human values (Kumar 2003). Jameson recognises the utopian impulse in everyday practice coming into being through the effects of absence. Accordingly:

> Its [Utopia's] function lies not in helping us to imagine a better future but rather in demonstrating our utter incapacity to imagine such a future – our imprisonment in a non-utopian present without historicity of futurity – so as to reveal the ideological closure of the system in which we are somehow trapped and confined.
>
> (Jameson 2004: 46)

With his understanding we would never be able to define the missing piece, but it nevertheless functions through keeping alive the possibility that a different world might be possible. The opportunity offered by utopian thinking, then, is not as a project of design and identification. It rather seeks to provide a practice or method for channelling the desires present for what a brighter and better future could be.

Utopian thought and utopia proper in climate change debates

There is a growing awareness that the climate change problem impacts differentially upon different segments of society – such as national governments, transnational companies and indigenous peoples – giving rise to a multitude of framings of climate change causation and appropriate response. In the words of Hulme (2009), we disagree about climate change because it can, at the same time, be a problem of market failure, technology, adaptation, lifestyle, overconsumption, exploitation of natural resources and global injustice, depending on one's perspective and context.

The messiness of climate change further aggravates our understanding of how to address it. Complexity is inherent, due to both the characteristics of the atmospheric system and the myriad activities that all affect atmospheric forcing. Understanding climate change thus provides a formidable challenge for scientists, policy-makers and the public alike. Nisbet (2009: 16) even noted that '[f]or many members of the public, climate change is likely to be the ultimate ambiguous situation given its complexity and perceived uncertainty'.

We argue that the significance and complexity of this problem provide an opportunity to explore and exploit the logic of utopian thinking. Being framed as a major challenge for society creates a need for thinking beyond business as usual, pushing us to think ambitiously about alternative futures. Climate change, further, becomes a valid element in contemporary futures thinking in general. This could include suggestions on how to achieve decarbonisation and images of the good life in a low-carbon future.

In climate science and policy, though, dystopian visions surface more frequently. Climate change is presented as a crisis, with social change to be motivated by society's need to respond in order to avoid the catastrophic impacts. Dystopian framing has popular purchase, as witnessed by films such as *The Day After Tomorrow*, but also in political statements like Al Gore's *An Inconvenient Truth* and scientific works, for example, in *The Stern Report* (Stern 2006). Given the failings of dystopian frameworks to promote concrete and timely action, can utopia offer an additional or alternative frame and motivation for social change?

In a review of policy documentation available through the UNFCCC documentation service we found only a few examples of the use today of utopian thought to inspire action, and none with the detail that might be described as moving towards a pronouncement of a utopian proper. Quite the reverse, as, for example, demonstrated by the constrained vision of the US Secretary of State for Energy, Steven Chu's, plans for a green energy future. While on a general level carrying a vague resemblance to utopian thought, plans are carefully expressed through concrete actions in the near-term future, avoiding detailed descriptions of what a green energy society would look like in the long term and so what more far-reaching goals and consequences of such a shift might be. Even with such a limited vision, conservative critics portray the plans as 'utopian', by which they mean unrealistic in a naive sense, and out of touch with technological or economic reality. Such short-term political visions are not uncommon and tend to have a restricted focus on a very particular, circumscribed topic (e.g. the hydrogen car), and are often also circumscribed by national interest and regulatory realism.

The following subsection presents a review of the contemporary framing of climate change according to the distinct logics of business/technological, indigenous and scientific utopias.

The business/technological utopia

Our examination of futures thinking in the business community found two types of utopian thinking. First, based on hopes for technological solutions as the pathway for achieving zero GHG emissions and, second, in support of alternative or slightly modified economic systems exploiting ecological modernisation and presented through terms including green capitalism, natural capitalism and reformed capitalism.

Technological innovations were frequently found to be advocated by business organisations in a utopian sense. In relation to climate change, several specific technical solutions were presented as principal parts in a zero- or low-carbon economy. These could be seen as embryos of technological utopias, providing a detailed description of one part of the low-carbon society. Alternatively, these confined technical visions could be argued to hinder thinking beyond business as usual and assuming no change in behaviour and masking alternative options to current Western lifestyles.

The futures of business organisations contain descriptions of only minimally modified economic systems connoted by calls for 'green', 'natural' or 'reformed' capitalism. One such example is the World Business Council for Sustainable Development (WBCSD 2010) 2050 vision report, which addressed three questions: What does a sustainable world look like? How can we realise it? How can business contribute to more rapid progress towards that world? Despite being a visioning document, the WBCSD were keen to distance the vision from utopian elements, stating: 'At first this Vision may read like a utopian ideal, considering how far it seems to be from the world of today. But that is neither the intention of this report, nor the reality' (WBCSD 2010: iii). However, the vision carries all the defining features of utopian thinking. Here is one example of how the vision is introduced. In 2050, people all over the globe are provided with the 'means to meet their basic human needs, including the need for dignified lives and meaningful roles in their communities' (WBCSD 2010: 6). An ideal of 'One World – People and Planet' permeates policies worldwide and encompasses an understanding of the interdependence among people and their dependence on the

Earth (WBCSD 2010: 6). This stated vision fits within a business worldview that is comfortable with the universalising tendency of globalisation, where goals and solutions are shared across all societies of the world. Conflicts, disasters and violence remain in this vision, but 'societies are resilient, able to withstand disruption and quickly recover' (WBCSD 2010: 6). It is a vision of a world society with high economic growth, but one decoupled from devastating ecosystems and excessive material consumption.

In our reading, WBCSD's 2050 vision portrays a classical utopia proper, since it contains a narrative describing a hypothetical society in full operation with desired features spelled out in great detail. The WBCSD report states that it 'does not offer a prescriptive plan or blueprint but provides a platform for dialogue, for asking questions' (WBCSD 2010: ii). Slightly upholding the current economic system, the vision elaborates on changes in values. Non-materialistic values are spreading and the vision anticipates unification around the goals and prescriptions for society.

Arcadian or indigenous utopia/dystopia

Indigenous peoples' voices provide an alternative debate to the dominant Western-oriented framing of global climate change (Smith 2007). We report here on an assessment of futures thinking as expressed in the statements delivered by indigenous peoples at UNFCCC COPs and in Bolivia's recent proposal for decision on a shared vision at the UN Climate Conference in Cancun (UNFCCC 2010).

Few indigenous voices offer utopian visions for the future. Instead the majority of formal declarations tend to be dystopian or apocalyptic, warning of a catastrophe if the global economic order is not dramatically changed. This is clearly illustrated by a quotation from the 1998 Albuquerque Declaration delivered at the UN Climate Conference in Buenos Aires:

> Our prophecies and teachings tell us that life on earth is in danger of coming to an end.... The destruction of the rest of the Creation must not continue, for if it does, Mother Earth will react in such a way that almost all people will suffer the end of life.
>
> (Albuquerque Declaration, article 1)

This is similar to Lovelock's proposition in *The revenge of Gaia* (2007). Such statements may though also be considered utopian in as much as they are highly critical of the current capitalistic system.

Advocacy for indigenous peoples made by the Chair of the Alliance of Small Island States also presented climate change as catastrophe, by stressing that the GHG emissions of Western, richer societies threaten current and traditional ways of life among the poor, including indigenous groups. From this perspective, climate change is even presented using the language of cultural and ecological genocide. Raising sea levels and temperatures have already affected culturally significant economic activities such as fishing, reindeer herding and other activities, and are likely to become even more pressing in the future. This view was restated recently by the Bolivian representative at the UN Climate Change Conference in Cancun, 2010, Pablo Solón Romero, who motivated Bolivia's persistence to block a final decision because it only allowed for a continuation of market-based climate policy, without the richer countries taking legal responsibility to reduce their emission, and did not provide the fundamental change towards a society which would avoid dangerous interference with the climate system, by commenting on the final agreement of the conference that:

> We're talking about a reduction in emissions of 13–16%, and what this means is an increase of more than 4C. Responsibly, we cannot go along with this – this would mean we went along with a situation that my president [Evo Morales] has termed 'ecocide and genocide.
>
> (BBC 2010)

In April 2010, the Universal Declaration of the Rights of Mother Earth was adopted by the World People's Conference on Climate Change. It also explicitly saw the dominant global capitalist system as the root of the current environmental crisis, and argued for a complete change of the economy and social organisation. The blame, thus, is put on:

> the capitalist system and all forms of depredation, exploitation, abuse and contamination have caused great destruction, degradation and disruption of Mother Earth, putting life as we know it today at risk through phenomena such as climate change.
>
> (Universal Declaration of the Rights of Mother Earth, preamble)

The Hague Declaration also stresses the inherent value of the Earth: the 'Earth is not a commodity, but a sacred space' (The Hague Declaration of the 2nd International Forum of Indigenous Peoples and Local Communities on Climate Change). This statement was repeated in the Bolivian proposal of a shared vision presented at the 2010 UN Climate Conference (UNFCCC, article 3). Both texts stress the non-materialistic and spiritual dimensions of nature, as well as offering a sharp criticism against the separation between human and natural systems or the idea that man is outside nature, i.e. the externalisation of the environment. These declarations do not suggest how the transition from the current economic and social model might unfold and in this sense also exhibit the freedom of utopian thought, but they call for a return to the Arcadian traditional ways of life.

From this short exposé, it is clear that indigenous positions tend to present dystopian frames. These statements contain a radical critique of the dominant global economic system and call for systemic transition to avoid catastrophe either for all mankind, or for traditional cultures, ecosystems or species. It also contains elements of Arcadian or ecological utopia.

Scientific utopias

The IPCC has served as a global focal point structuring and synthesising scientific observations and findings on climate change since its establishment in 1988. The IPCC's mandate is to assess 'the scientific, technical and socioeconomic information relevant for the understanding of the risk of human-induced climate change'. To provide a robust answer of what would be the likely status of temperature, precipitation and wind in the future, general circulation models are employed to generate climate change scenarios. Socio-economic scenarios have been vital to guide the economic and energy modelling needed to estimate the amount of GHGs emitted by the activities in a range of future societies (Swart *et al.* 2004). In this sense, climate change research has had a relatively long tradition of thinking about social futures (Nakicenovic and Swart 2000; Moss *et al.* 2010). Already in the early 1990s, the IPCC had decided to develop new socio-economic futures due to the dramatic geopolitical change when the Cold War ended (Nakicenovic and Swart 2000).

In a previous study we demonstrated that utopian thought had surfaced in climate change science only as fragments through the construction of the integrated emissions 'SRES' scenarios (Hjerpe and Linnér 2009). The SRES scenarios contain social futures visioning, but these are tightly constrained by assumption and methodology, for example, through being forced to assume global decoupling of GHG emissions from economic growth, an even geographical and social distribution of the benefits of economic globalisation, and of technological development.

This said, the resultant SRES socio-economic scenarios do all depart marginally from the current economic order. Ranges of futures visioning are constructed through divergent values surfaces through contrasting scenarios based on business as usual or more environmentally sensitive economies, as well as on the relative role of the state or market in driving technological innovation. Scale is considered too in distinguishing between globalisation trends on the one hand and regionalisation on the other. This also has a value dimension expressed through preference for a global free trade world, a locally oriented small-scale economy world or a global sustainability-oriented world. The scenarios are derived from the interaction of only two dimensions in any one model, leading critics to argue that the SRES 'range' of scenarios is too narrow (Carlsen and Dreborg 2008; Raskin *et al.* 2010). Partly in response to this, the IPCC invested in a second generation of socio-economic scenarios labelled Representative Concentrations Pathways (RCP) to be developed (Moss *et al.* 2008). These, perhaps inadvertently, provide more scope for guided utopian thought in three ways. First, by dividing scenarios into short- and long-term visions they reduce tension between planning and a preference for short over long termism. There will be outlook scenarios for 2300. Here there is a high degree of freedom for thinking about alternative futures because in that far time perspective no future vision can claim authority over others. Second, by emphasising GHG concentration pathways, there is more room for pluralism in storylines and eventual scenarios as many alternative futures may lead to the same trajectory or to the same atmospheric concentration. Third, scenarios will contain the effects of climate policy. Since the SRES scenarios did not consider climate change policy, this opens up possibilities for different assumptions of policy responses in the futures scenarios. Hence, it establishes a platform to elaborate the kinds of climate change policies particular future societies should and could implement. Furthermore, when the RCPs were selected, policy-makers pushed for a very low GHG pathway – a pathway that at the time was not replicated by other models as is usually the case for being a RCP candidate – involving negative GHG emissions in the intermediate future. This clearly contains a utopian element and significantly breaks with current political trends.

Through the process of scientific evolution, the objects of climate change science are also increasing with consequences for those elements of life that are included in potentially utopian visions. While scientific, the debate hitherto has mainly focused on political and technical solutions, and the latest synthesis report of the IPCC (2007) emphasises the importance of changes in lifestyle and behaviour patterns. The IPCC Chair, Dr Rajendra Pachauri, has increasingly urged for incorporation of lifestyles in social climate science and policy. He has called for vegetarianism and aviation taxes. These are signs that lifestyles are becoming negotiable and the everyday has become an object of policy, with the potential to bridge the gap between utopia and practice.

What could utopia proper offer political action?

We have found few examples in the policy and scientific literature of utopian thought for action, and none deploying utopia proper. The quietness of utopian thinking in the UNFCCC process can serve as an illustration of the difficulties of applying utopian thinking in international climate politics more generally. In spite of plentiful references to long-term visions of low-carbon futures, the future is rarely envisioned in its potentially rich diversity, nor in detail, save through the vision of a prosperous, decarbonised society driven by solar, wind and often nuclear energy, and fully commercialised carbon capture and storage. The fragments of utopia embedded in assumptions of rates for various IPCC socio-economic modelling parameters run the risk of being masked as predictions or understood as preferable or inevitable. But these fragments do not provide us with any picture of the foundation on which these economic, social and technological relations rests and as such they are not intended to indicate preferences for the future, although in making judgements on the likely range of future worlds they do contribute to future framing for development discourse.

As our study suggests, utopian thinking is marginalised in contemporary climate debates and if it surfaces it is primarily among non-state actors presenting alternative futures. The reason is not the capitalist logic itself, but rather a combination of the main purpose for most of the futures thinking. The frequent use of scenarios in climate change science in general and the implications of the new RCP scenario approach in particular may provide an avenue for utopian thinking. However, the existing SRES narratives do not contain elements of utopia proper and in this way eventually serve to limit visions of the future that are considered by policy-makers (Raskin *et al.* 2010).

In serving as analytical tools, these future visions have been useful, but to stimulate thinking about alternative futures the usefulness of the SRES scenarios has been very limited, since their focus has entirely been on envisaging how global warming may evolve under different extrapolations of socio-economic trends, but they rarely consider how different decarbonised futures should be organised. We suggest that the RCP approach is promising in that it may open up for utopia proper. It brings in the very long-term view and since no vision of society beyond 2100 could truthfully be claimed as more or less plausible as another, radical futures could also be elaborated. The scenarios also enable formulations of what climate change policies the various good societies should and could implement. Whether this happens is up to climate scientists and the informed public.

To capture fully the potential of utopian thinking, socio-economic scenarios need to be spelled out in more detail. If this is to be done, climate science has to bite the bullets of subjectivity and positionality and accept the difficulty of revealing the meaning, basis and limits of the scenarios without attaching preferences to them (Holman *et al.* 2005). Elaborating details means answering key questions on what is good and bad in society, and prescribing social organisation and modes of economic production and exchange. This process can open up valuable political space for dialogue – details are necessary to confront society with what may be deemed as desirable and may need to be adjusted or discarded to move closer to a utopian ideal. Current, vague, utopian-like fragments, such as calls for a low-carbon future with global economic convergence, are attractive to all but say little about how sustainability may be reached and the trade-offs required in arriving at a more sustainable future.

Across the climate change literature, dystopian thought was common, particularly in indigenous peoples' declarations and proposals. Serving as an important reminder of the consequences of overuse of the atmosphere as a GHG sink, they are evoked to push for a radical transition of the current economic system to avoid a catastrophe. These visions were also marshalled to argue for alternative models for climate change mitigation and limited or no room for carbon markets, but with no explanation for how such a transition might unfold. Rather, alternatives tended to rely on a return to or continuation of the romanticised, historical and sometimes indigenous ways of life. This emphasises the current constrained imagination and lack of diversity and imagination in thinking about alternative futures. Underlying this is perhaps a pragmatic concern that futures calling for radical change – especially in everyday activities – will be challenging and likely unpopular at first glance. In contrast to this lack of visioning for utopian alternatives from those most exposed to the impacts of climate change, the business community presented futures including a hypothetical society in full operation with desired features described. This was a vision supportive of zero-carbon economies and low-carbon technologies but with no references to the economic system or dominating values that might achieve and maintain such change.

Our reading of future visions in various policy documents not surprisingly suggests very different roles for the state and political space. Business-centred visions stressed the principle of subsidiarity and coordination between the state and non-state actors as well as between other states. Indigenous peoples' future visions emphasised community and internalised the environment, suggesting a wide basis for mitigating climate change, even though specific measures may be restricted by cultural norms. In the SRES socio-economic scenarios the role of the state differed, but each scenario departed from current ideas of liberal democracy and free market capitalism. The RCP approach opens up far more scope for the presentation of utopias proper, with specific visions of future societies are not prescribed. In this way, the room for climate action and the role of the state can be differentiated across alternative futures.

If we are to move away from mere resilience of current society and actually develop decarbonised societies in line with the ideals of global sustainable development, having inter- and intra-generational equity as key criteria, planning tools will be needed. These planning tools should help us to recognise and develop new alternative options and pathways. Or rather, in line with Jameson (2004), planning tools help us to realise our difficulty to imagining in detail how future good societies may be organised and how entrapped our creativity is in current political, economic, social and technical contexts.

To transform societies into the bold goal of a sustainable development capable of phasing out our GHG dependency while enhancing quality of life for all in line with the UN Millennium Development Goals, we need radical and systematic changes in our means of production and consumption. We need to move beyond resilience, away from lengthy negotiations of trying to reactively regulate problems with minor adjustments of contemporary ways of managing society. In this chapter we have argued that and exemplified how utopian thinking could spur debate and creativity on how systemic change may be achieved. Our exposé has also shown a lack of utopia proper in policy and scientific discourse, and identified potential entry points for debate and analysis of alternative futures using a utopian frame.

References

BBC (2010) www.bbc.co.uk/news/science-environment-11975470.

Block, E. ([1918] 1923) *Geist der Utopie*, Berlin: Paul Cassirer.

Carlsen, H. and Dreborg, K.-H. (2008) Dynamic generation of socio-economic scenarios for climate change adaptation: methods, building blocks and examples (in Swedish), FOI-R-2512-SE.

el-Ojeili, C. and Hayden, P. (2006) *Critical Theories of Globalization* London: Palgrave Macmillan.

Giddens, A. (2009) *The Politics of Climate Change*, Cambridge: Polity Press.

Hardt, M. and Negri, A. (2000) *Empire*, Cambridge, MA: Harvard University Press.

Hedrén, J. and Linnér, B.-O. (2009) Utopian thought and the politics of sustainable development. *Futures* 41(4): 210–219.

Hjerpe, M. and Linnér, B.-O. (2009) Utopian and dystopian thought in climate change science and policy. *Futures* 41(4): 234–245.

Holman, I.P., Rounsewell, M.D.A., Shackley, S., Harrison, P.A., Nicholls, R.J., Berry, P.M. and Audsley, E. (2005) A regional, multi-sectoral and integrated assessment of the impacts of climate and socio-economic change in the UK. *Climatic Change* 71: 9–41.

Hulme, M. (2009) *Why we Disagree about Climate Change: Understanding Controversy, Inaction and Opportunity*, Cambridge: Cambridge University Press.

IPCC (2007) *Contribution of Working Group III to the Fourth Assessment Report of the Intergovernmental Panel on Climate Change, Summary for Policymakers*, Cambridge: Cambridge University Press.

Jameson, F. (2004) The Politics of Utopia. *New Left Review* 25: 35–54.

Jorgensen, D.J. (2005) The indigenous utopian. Paper presented at Imagining the Future: Utopia, Dystopia and Science Fiction Conference, Monash University, 6–7 December.

Kumar, K. (1987) *Utopia and Anti-Utopia in Modern Times* Oxford: Blackwell.

Kumar, K. (1999) *Utopianism*, Milton Keynes, UK: Open University Press.

Kumar, K. (2003) Aspects of the western utopian tradition. *History of the Human Sciences* 16(1): 63–77.

Lovelock, J. (2007) *The Revenge of Gaia. Earth's Crisis and the Fate of Humanity.* London: Penguin Books.

Moss, R., Babiker, M. and Brinkman, S. (2008) *Towards New Scenarios for Analysis of Emissions, Climate Change, Impacts, and Response Strategies. Technical Summary*, Geneva, Switzerland: IPCC.

Moss, R., Edmonds, J. and Hibbard, K. (2010) The next generation of scenarios for climate change research and assessment. *Nature* 463: 747–756.

Nakicenovic, N. and Swart, R. (eds) (2000) *Emissions Scenarios, Summary for Policymakers: A Special Report of IPCC Working Group III*, Cambridge: Cambridge University Press.

Nisbet, M.C. (2009) Communicating climate change. Why frames matter for public engagement. *Environment* 51(2): 12–23.

Prins, G., Galiana, I. and Green, C. (2010) The Hartwell Paper: a new direction for climate policy after the crash of 2009. Institute for Science, Innovation & Society, University of Oxford, London School of Economics and Political Science, London.

Raskin, P.D., Electris, C. and Rosen, R.A. (2010) The century ahead: searching for sustainability. *Sustainability* 2: 2626–2651.

Smith, H. (2007) Disrupting the global discourse of climate change: the case of indigenous voices, in Pettenger, M. (ed.) *The Social Construction of Climate Change: Power, Knowledge, Norms, Discourses*, Ashgate: Aldershot (pp. 197–215).

Stern, N. (2006) *Stern Review on the Economics of Climate Change.* HM Treasury Review.

Swart, R.J., Raskin, P. and Robinson, J. (2004) The problem of the future: sustainability science and scenario analysis. *Global Environmental Change* 14: 137–146.

UNEP (2010) The Emissions Gap Report. Are the Copenhagen Accord pledges sufficient to limit global warming to 2 °C or 1.5 °C? A preliminary assessment.

UNFCCC (2009) Copenhagen Accord.

UNFCCC (2010) Proposal on draft decisions submitted by the Plurinational State of Bolivia. FCCC/AWGLCA/2010/CRP.4.

Universal Declaration on the Rights of Mother Earth, available at: http://climateandcapitalism.com/?p=2268.

Wallerstein, I. (1986) Marxisms and utopias: evolving ideologies. *The American Journal of Sociology* 91(6): 1295–1308.

World Business Council for Sustainable Development (WBCSD) (2010) *Vision 2050: The New Agenda for Business.*

12 Resource exchange, political strategy and the 'new' politics of climate change

Ian Bailey and Hugh Compston

Introduction

It has become almost obligatory in recent decades for international conferences to begin with plenary declarations by national leaders about their country's commitment to achieving a low-carbon economy. Despite this, progress towards large-scale reductions in greenhouse gas emissions remains faltering in most countries. Some of the reasons for this relate to the inherent uncertainties of predictive environmental science, while others are indicative of technology and economic constraints (IPCC 2007; Stern 2007). The most intractable problems, however, appear to be political: most leaders are concerned about environmental change and have introduced emissions-reduction measures of some description. But they are also aware that climate-related policies which seriously disturb established economic and social practices are likely to trigger resistance from business groups, electorates, parliaments or even their own parties that may lead to irreversible damage to their party and their own political careers (Compston and Bailey 2008). Faced with this predicament, most national climate programmes have erred towards concrete short-term and aspirational long-term targets, the use of market-based measures to encourage clean technologies, and calls for more international cooperation, but the singular avoidance or dilution of measures that might lead to serious losses of political capital for the government.

Optimists might suggest that mounting scientific and public anxiety about climate change will combine with growing policy experience to corrode political obstacles to more ambitious climate policy without the need for a major rethink of political tactics. Governments may even be able to enhance their political capital if they can establish a strong constituency of support for climate policy through appeals to public and business self-interests and altruistic concerns, or even through questioning some of the conventional assumptions of capitalism. A more likely situation is that entrenched interests will not be dislodged easily, that current approaches will face diminishing returns as 'no-regrets' options dwindle, and that new forms of climate politics will be needed to manage tensions between the demands of consumer capitalism and global environmental change (Boykoff 2009; Giddens 2009; Giddens *et al.* 2009; Hale 2010). The systemic nature of the climate challenge suggests that such a new politics would need to move beyond narrow debates on the selection and design of policy instruments whose relationships with the structural causes of climate change are often haphazard towards greater innovation in how governments gain the support or acceptance of potential veto groups, or at least counter their capacity to inflict political damage on the government (Boykoff 2009; Bulkeley and Newell 2010).

In this chapter we review some of the relational and systemic challenges involved in creating such a new politics of climate change. We begin by reviewing the current literature on the topic, where we argue that analysis needs to move beyond descriptive-theoretical and critical accounts of existing climate politics, and beyond visionary but abstract calls for new social and economic values, to engage more explicitly with the 'day-to-day' practicalities of reforming institutional structures and processes. We then examine how perspectives based on resource exchanges within climate-policy networks may assist in identifying political strategies to limit the political risks of strengthening climate policy, and outline a series of resulting strategies. We conclude by offering conclusions on the utility of resource exchange as a lens for defining new political strategies for climate policy.

The politics of climate change

In his recent book *The Politics of Climate Change*, Anthony Giddens (2009: 4) claims that '*we have no politics of climate change*' (emphasis in original); or to put it less theatrically, academia and politicians have yet to produce a well-developed analysis of the political innovations needed to connect supranational and state responsibilities, party politics, technologies, markets and civil society in a coherent way to effect a transition to a low-carbon economy. Continuing this tack, Giddens reproaches Lord Stern (2007, 2009) for failing to consider the politics underlying strategies to reduce greenhouse gas emissions:

> Extraordinarily, there is no mention of politics in Stern's discussion, no analysis of power. It is as if the 'global deal' will be reached as soon as the nations of the world see reason.
>
> (Giddens 2009: 201)

A brief examination of the literature suggests that Giddens' remarks do an injustice to the accumulating body of scholarship on the politics of climate change. For example, the ADAM (Adaptation And Mitigation) project, a consortium of 24 European and international research institutes. recently completed an investigation of the political dilemmas facing the European Union (EU) in respect of climate governance (Jordan *et al.* 2010). Equally discounted is the literature on the hybridisation and deterritorialisation of climate politics resulting from the activities of multilateral bodies such as the WTO and climate governance networks, like the *Cities for Climate Protection*, that operate within and across national boundaries (Bulkeley 2005; Biermann and Pattberg 2008; Pattberg and Stripple 2008; Bulkeley and Newell 2010). Also apparently neglected are the numerous studies examining the governance challenges involved in the creation and management of international carbon markets such as the EU emissions trading scheme, Kyoto flexibility mechanisms and voluntary carbon offsets (Bailey 2007; Skjærseth and Wettestad 2008; Boyd 2009).

There nevertheless remains some truth in the claim that surprisingly few writings on the politics of climate change are dedicated to exploring practical strategies for unblocking political obstacles to stronger climate policy. Most focus instead on descriptive-theoretical and critical accounts of current climate-governing arrangements, including: critiquing the performance of national climate strategies against claims about their effectiveness made by politicians (Kerr 2007); analysis of the environmental and equity outcomes of climate initiatives associated with international carbon markets (Lohmann 2005; Newell 2009; Paulsson 2009; Newell and Paterson 2010); and examining the contested politics of translating mitigation and adaptation goals from the international to the national and local spheres (Bulkeley 2005; While *et al.* 2010).

A second category, meanwhile, involves normative accounts of the reforms their authors see as necessary to catalyse a more progressive climate politics. Hale (2010), for example, contends that conventional politics is structurally incapable of liberating itself from its allegiances to global capital and that the third sector should lead a grassroots mobilisation to pressurise those in political authority to take climate change seriously. In a similar vein, Pendleton (2010) argues for a new goal of a climate-compatible political economy as a precursor to international agreement and national actions, while Bailey and Wilson (2009) and Castree (2010), among others, critique the neoliberal ideologies underpinning global carbon markets and argue for more reflexive transitional processes.

While these studies form a legitimate part of critical reflections on whether established economic, social and political wisdoms have the capacity to manage the risks of climate change, they lack a clear methodology for *how* to proceed away from the status quo. Even Giddens' (2009) own attempts to identify principles for a new climate politics suffer from something of a detail deficit. For example, among the concepts Giddens advances is that of the *ensuring* state:

> The enabling state suggests that the role of the state is confined to stimulating others to action and then letting them get on with it. The ensuring state is an enabling state, but one that is expected or obligated to make sure such processes achieve certain defined outcomes – in the case of climate change the bottom line is meeting set targets for emissions reductions.
>
> (Giddens 2009: 9)

But what steps are involved in moving from an enabling to an ensuring politics? Who might oppose it and how might opposition be counteracted? Other suggestions by Giddens are equally laudable but unstipulated. For instance, how *do* politicians: aid capacity to think ahead; promote political and economic convergence; integrate policy at different spatial scales; recognise the development imperative for poorer nations; institutionalise the polluter-pays principle; counter business interests that seek to block climate initiatives; or develop an economic and fiscal framework for moving towards a low-carbon society? Giddens elaborates how some might be advanced, but often, probably out of necessity at the enormity of the task, he stops short of systematic examination of the political tactics involved in achieving each end.

Despite growing attention to the political dimensions of climate policy, therefore, the literature has yet to define a clear a set of practical strategies to tackle political obstacles that have contributed to the sclerotic nature of climate politics in many countries. In the following section, we outline a resource-exchange perspective of climate-policy networks in order to explore how recognition of mutual resource dependencies between state and non-state actors might be utilised to identify and interrogate political strategies for reducing these obstacles. We begin by reviewing the main resource interdependencies in such networks before examining political strategies for climate policy that may be drawn from the concept of resource exchange.

Policy networks, resource exchange and political strategies in climate policy

The term 'policy network' has been employed by various disciplines to describe and explain the processes by which policies are agreed and implemented. In its most basic sense, it refers to sets of political actors inside and outside government who are involved in – or have some interest and influence over – public policy, and to relations between these actors. Börzel (1998: 254) notes that, despite variations in the use of the network concept:

> They share a common understanding, a minimal or lowest common denominator definition of a policy network, as a set of relatively stable relationships which are ... interdependent linking a variety of actors, who share [an interest] with regard to a policy and who exchange resources to pursue these interests.

In simple terms, interactions between network participants are based on resource interdependencies: each actor has preferred outcomes, possesses some of the resources needed to formulate and/or implement policy, and may exchange resources to promote their interests.

The first points to note about climate policy networks are: (1) that the structural changes needed to achieve step reductions in greenhouse gas emissions cannot be organised unilaterally by government; and (2) that the changes implied by a low-carbon transition do not align neatly with the existing preferences of the main actors whose cooperation is needed. Politicians and administrators will have preferred policies and outcomes and will additionally seek to uphold and strengthen their departments. Industry may offer conditional support to manage current and future climate-related risks but will also seek to defend other commercial interests (Gouldson and Bebbington 2007). Environmental NGOs tend to want stronger climate policies, while voters may desire action on climate change but be reluctant to forgo the benefits of high-carbon lifestyles. As such, state and non-state actors are in a resource-interdependent situation and will use political strategies to promote their preferences while seeking to surrender as few resources as possible to avoid weakening future bargaining positions (Rhodes 1985; van Waarden 1992; Klijn and Koppenjan 2000; Thompson 2006). The main tradable resources held by each actor group are summarised in Table 12.1.

Five main forms of resource-exchange strategies that governments might use – or use more vigorously – to strengthen climate policy while avoiding significant political damage are now discussed (Compston 2009).

Unilateral decision-making

Instinctively, unilateral action is inconsistent with avoiding political damage, since it implies overriding objections to new policies and taking a calculated risk about the consequences. Governments may, of course, avoid this problem by only proposing policies on which all major actors agree but this is unlikely to produce significant deviations from business-as-usual emissions trajectories. They may also cultivate alternative sources of support by taking a stand against big capital but are still likely to face attempts to obstruct policy or undermine the government. In resource-exchange terms this means that governments risk greater political damage and losing access to key resources, such as continued investment by major companies or the political support of electorates. There are, of course, exceptions, such as introducing policies to which no major party objects and making incremental changes that progressively strengthen policy. However, the prospects for this depend on the government's skill in breaking ambitious policies into increments that do not provoke cumulative opposition while maintaining a clear, long-term policy direction. In addition, the limited dialogue implied by unilateral action may cause other actors to withhold information needed for the design of effective and cost-efficient policies.

The three main tradable resources that governments cannot survive without are political support, cooperation with implementation and private investment. Loss of political support can lead to loss of office – via a leadership challenge, confidence vote or election[1] – and the possible reversal of new climate policies, so the risks taken are in vain, both from a political and a policy perspective. Clear legal powers and strong sanctions are required to avoid non-cooperation with implementation undermining the credibility of policy, while losses in private investment (short-term capital flight or loss of new investment) may cause economic problems that damage the government's re-election prospects.

Unilateral approaches thus hinge on the governments' ability to devise tactics to limit the political damage caused by losing these tradable resources. Options here include:

- Introducing contentious policies early on during an administration to allow opposition to subside and the benefits of policy to become clearer before the next election.

Table 12.1 Main tradable resources of policy-network members

Controlled by	Resource	Description
Public actors alone	Policy amendments	Changes in policy instruments or settings. Only actors with the legal authority to make binding decisions can trade policy amendments. Policy amendments may be traded between branches of governments.
	Access to decision-making	For example, contact with officials or politicians, inclusion on committees, invitations to contribute to consultations. Access gives non-state actors information on government policy plus the chance to present arguments. It may be mandated or be granted by public actors.
Public and private actors	Veto power	For example, obstruction of policies by opposition parties unless amendments are made. The tradable resource consists of refraining from exercising veto power.
	Information	For example, exchange of privileged information for policy amendments. Information may also be used to change the preferences of public actors, including through promoting policy learning.
	Cooperation with implementation	Where actors are able to hinder implementation legally, public actors may exchange amendments for cooperation with implementation.
	Recourse to the courts	Where public or private actors are able to use legal proceedings to block a policy, refraining from using this option can be traded for policy amendments.
	Political support	Private actors may mobilise the public or groups for or against a policy. Support from legislative bodies, the political party and the head of government are all important. Parties outside government may also try to trade political support for the government for policy amendments. Politicians may deal directly with voters by amending policy in exchange for opinion-poll ratings. The significance of political support depends on how much the public actor needs it and on perceptions (e.g. whether environmentalists can mobilise voters is uncertain ahead of being demonstrated).
	Patronage	For example, public actors may trade positions linked to government for investment or campaign donations. Private actors may offer jobs to ex-public servants in exchange for policy amendments while in office.
Private actors alone	Private investment	For example, withdrawal, continuation or expansion of private investment by companies in exchange for policy amendments. Again, threats to disinvest are only effective if the government believes they are credible.
	Fluid funds	For example, bribes, campaign contributions, buying expertise, lobbying services and other resources.

- Targeting measures on a smaller range of industries to reduce the number of potential opponents the administration has to manage. How far governments can do this will, of course, be limited by equity norms and avoiding hostility among sectors that fear being targeted next.
- Targeting economic sectors that are most able to pass on additional costs to consumers, since this may spread burdens and facilitate the internalisation of environmental costs without government being blamed directly, especially where market forces allocate costs in ways that avoid major injustices. Even assuming this is possible, media coverage of the effects of taxes on consumer prices, for example, may aggravate political damage.
- Adopting policies that target losses on groups least able to inflict political damage via the ballot-box, implementation or investment. Such an approach may be ethically suspect, however, unless a clear justification can be made against the greenhouse gas benefits of such an approach. Even then, it may be politically damaging if injustices are publicised in the media.

Simple resource exchange within existing network parameters

The most obvious way for governments to avoid political damage in climate policy is to trade policy amendments for resources while attempting to avoid being over-generous. Strategies of resource exchange under these conditions are thus likely to consist of: (1) each actor communicating its preferences and arguments; and (2) deciding on the types and sequencing of resource exchanges. The actions of each actor are, of course, influenced by those of others: players 'gear their actions and the objectives which they pursue to the strategic behaviour and objectives of other actors' (Klijn *et al.* 1995: 440).

Governments face two major choices within this framework. They must decide whose support is essential, as in general the greater the number of actors whose agreement is sought, the more policy concessions the government is likely to have to offer (Nunan 1999). They must also decide what changes they can make without substantial policy drift from its original objectives. This requires a priori judgements about which amendments will pacify business and electoral opposition and which elements are indispensable to the policy. If the balance between what is demanded and what government can offer is seen as unacceptable, it may take a calculated risk on unilateral action.

Where amendments are made, these can relate to the climate policy under discussion (e.g. the 80 per cent reduction in the UK's climate change levy (CCL) offered to energy-intensive sectors in 2001 in exchange for legally binding agreements to reduce emissions (Bailey and Rupp 2005), or as part of a package of policies affecting the same actors. For instance, industry may consider trading climate policies for concessions in taxation, business regulation and labour law, as was again the case with the CCL, where the government introduced a 0.3 per cent reduction in employers' national insurance contributions.[2] Macdonald (2008: 239) proposes a similar approach in Canada, where opposition from the

Albertan provincial government has impeded the agreement of federal climate policies. Macdonald advocates the granting of transitional subsidies to Alberta to assist with industrial restructuring away from oil-related industries while sending 'a tough-love message' that the national government 'will assist, but will no longer turn a blind eye to inaction'.

Facilitating agreement by facilitating resource exchange

As well as working within existing network parameters, governments may manage policy networks in order to facilitate resource exchange. Options here include: nurturing new lines of communication and working with actors whose participation is essential to a particular task to build mutual understanding and trust; orchestrating interaction and formalising the rules that regulate interaction, for example, by establishing conflict-regulating mechanisms; matching problems, solutions and actors; and the development of mediation and arbitration procedures (Kickert and Koppenjan 1997). The most relevant of these to climate policy seems to be the creation of procedures that facilitate interaction through committees, networking events and establishing norms of interaction (Jordan *et al.* 2010).

Changing the preferences of other actors

A fourth set of strategies involves changing other actors' preferences by altering their perceptions of 'the problem', solutions, the policy process, or the pressure they face from public opinion or investors for continued obstruction. The media are likely to be central to such efforts, since almost all political communication targeting mass audiences is mediated through media coverage (Gavin 2009). The most evident communication strategy is the provision of information on the biophysical, economic and social threats of climate change, and practical responses that might be developed. Another is to stress the co-benefits of climate policy, such as improved energy security and employment. In addition to accuracy, clarity and simplicity, deploying metaphors and analogies may help to make complex ideas accessible to the intended audience. Social marketing tailored to particular audiences (e.g. social groupings) might assist, as may frequent repetition, though one must be careful to avoid creating audience fatigue (DEFRA 2008).

Related to these is the idea that political actors use narrative devices, such as plot and characterisation, as well as evidence and logic, to secure support for their definition of reality (Hajer 1995). Recent innovations here include promises of a Green New Deal, while in the USA the Apollo Project likens the task of controlling climate change to the effort during the 1960s to launch a manned mission to the moon, though none of these representations are unproblematic (Pralle 2009). Increased incidences of climate- and fossil-fuel-related events (Hurricane Katrina and the Gulf of Mexico oil spill) may also energise the public's appetite for climate policy, although episodes like the Climate-gate scandal equally reveal the scope for media coverage and public sympathies to move in the opposite direction. The unpredictability of events and media coverage notwithstanding, their strategic significance lies in creating windows of opportunity when governments may create new policy agendas at lower political cost. Governments must, however, be wary of accusations of knee-jerk policy-making (Niemeyer *et al.* 2004), even when they have fully prepared policy options ready for spikes in public concern about climate change.

Moves to alter the rules, norms and terms of resource exchange

Five main options exist for governments to change the balance of power between themselves and other actors in respect of climate policy:

1 *Reduce the government's concern about whether certain resources are forthcoming from other actors:* for example, by switching to policy goals that require fewer externally held resources. Governments may avoid legislative veto points by prioritising climate policies that do not require legislative approval, such as some regulatory instruments or changes in instrument settings. This strategy is already being used to a high extent, for example, through the inflation indexing of the UK's climate change levy since 2007.

2 *Cultivate alternative sources for obtaining political support, help with implementation and business investment:* new sections of the electorate might be nurtured to compensate for losses in electoral support elsewhere, while policies that would fail because of non-cooperation with implementation may be replaced by policies for which cooperation is more forthcoming (e.g. German debates during the 1980s and 1990s on carbon taxes and industry self-commitments (Bailey and Rupp 2005)).

3 *Encourage other actors to increase their investment in resources that government controls:* one possibility noted earlier is to offer policy concessions elsewhere (e.g. taxation) for cooperation on climate policy. Another is to acquire additional powers to coerce other actors. The creation of the UK's Infrastructure Planning Commission in 2009, for example, increased the government's resources over planning decisions, enabling it to operate more freely than under a more localised planning system in expanding the renewable-energy sector. Another possibility is to take public ownership or greater control of selected firms.[3] Strong-handed state planning fell out of favour in many countries with the roll-out of neoliberal reforms during the 1990s, but has been debated and acted upon more vigorously since the global economic crisis (Giddens 2009). One example was the UK government's taking ownership of key financial institutions during the 2008 financial crisis. Some parallels may exist with energy production if climate change is interpreted as a sufficient emergency that its prompt decarbonisation becomes non-negotiable rather than a matter for designer markets. More moderately, extensions of regulatory powers and direct investment may be attempted.

4   *Deny other actors alternative ways of directing resources:* this may involve new controls on international transactions to prevent firms from shifting investment overseas. For instance, the EU considered introducing border-tax adjustments for imports from countries that do not have a comparable carbon price to that operating in the EU, although this initiative stalled when the EU Trade Commissioner, Peter Mandelson, suggested that it may breach WTO rules. A further possibility is to nurture cross-party consensus, such as that established for the UK's Climate Change Act, to limit the scope for business groups or voters to shift political support.

5   *Alter the structure of policy networks:* for example, by giving new actors access to policy-making – through consultations and committees – while excluding others. In some cases, this may make newly included actors more cooperative, although it may rebound if they use their influence to obstruct initiatives and/or alter the attitudes of previously cooperative actors. While excluding disruptive actors may increase their opposition, it also removes the need to include their preferences in policy. Second, governments may seek to establish or change organisational arrangements, such as consultation procedures and advisory bodies. In resource-dependency terms, this means altering policy network rules and norms, although it is uncertain whether or how this may affect the chances of actors agreeing to policies they previously rejected. Third, public actors may improve information systems, give formal recognition to an organisation (e.g. an industry body or NGO) as an interlocutor, grant access to permanent consultative bodies, provide subsidies, or grant a legal monopoly to elicit the cooperation of actors that receive these benefits. Fourth, public actors may alter which public bodies hold formal decision power in a particular area. Merging ministerial portfolios and moving responsibility for energy from an economic to an environment ministry, as has occurred with the creation of the UK Department of Energy and Climate Change, theoretically shifts power towards actors who may favour stronger action. Conversely, it may give opponents greater access to areas of climate decision-making that were barred by previous governmental structures.

Conclusions

We began this chapter by arguing that many of the root problems facing climate policy stem from a deeply held – and justified – fear among politicians that attempts to introduce radical emissions-reduction measures would meet fierce resistance from influential groups who see their interests being compromised, leading to political damage to the policies, politicians and governments involved. Scientific uncertainty, technology constraints and economic considerations all feature in these calculuses, and when combined with the logics of political survival may betray more structural misalignments between climate policy and the goals of contemporary capitalism. Either way, whether 'politics as usual' has the capabilities to manage the challenges of environmental change, rather than merely postponing the unravelling of capitalism's ecological contradictions, remains a very open question.

Despite a growing literature on the politics of climate change, academics have yet to produce a detailed analysis of political strategies to ease the introduction of policies capable of limiting climate change to acceptable levels. In this chapter, we have sought to further this debate by sketching the outlines of how a resource-exchange perspective on climate policy may help to provide governments with a suite of political strategies to lessen the political repercussions of pursuing stronger climate policies.

Having explored the potential of this approach, it is appropriate to conclude with some critical reflections. First, our general description of political strategy options provides few clues as to which resource exchanges are best suited to overcoming specific sources of opposition to specific measures in specific contexts; greater attention to such details is vital to identify the conditions under which individual strategies are more or less appropriate and how they might operate. A second omission concerns issues of scale and place in climate governance. Although rigid notions of discrete and hierarchical scales of governance are increasingly seen as untenable for interrogating multi-scalar and multi-issue problems such as climate change, the siting of renewable-energy projects and incentivising lower-carbon commuter zones have clear regional and local governance dimensions and may encounter materially different political obstacles to those influencing national climate policy and politics (Anable and Shaw 2007). As While *et al.* (2010) note, carbon control at the urban and regional scales is unlikely to consist of a simple cascade of national requirements. Rather, it will involve geographically contingent conflicts, power struggles and strategic selectivities as local governments try to reconcile environmental protection with other pressures and demands.

Third, our analysis has not probed the ramifications of combining political strategies. For instance, *spill-over policies* that create political and functional momentum for new or stronger policies may assist in breaking major reforms into less controversial and burdensome segments (Compston and Bailey 2008). This approach has been fundamental to the *Monnet method* of integration that transformed the EU from a customs union to a single currency, and the development of its environmental policies. However, spill-over strategies combine elements of resource exchange within existing parameters, changing the terms of exchange, and altering actors' preferences through norm-building, so defy simple categorisation. In reality, multi-strategy approaches are likely to be the norm but inhibit analysis of the effects of each intervention.

Two final reservations relate to whether the political strategies approach represents a genuine departure from – or simply window-dressing for – politics and economics as usual. After all, political strategising is a political 'stock-in-trade', so politicians may learn little from this analysis. What matters more, perhaps, is investigation of how to build political will rather than trying to teach politicians about tactics. However, few governments have made major breakthroughs in reducing greenhouse gas emissions and fear of political damage has been a major – if not *the* major – cause of this. This suggests the need for deeper analysis of how resource-exchange strategies can reduce the political risks created by climate policy. Equally, working with the political grain rather than challenging it might appear anti-radical and condoning of political-economic practices that have created – or at least contributed to – environmental and economic crises. However, political strategies that deal with current circumstances are more productively viewed as exploring pathways that may lead to more radical reforms. A final and perhaps decisive argument is that approaches which seek to reform institutional processes rather than trying to reinvent political culture in order then to find ways to avert further environmental-economic crises stand a greater chance of finding their way into the political mainstream sooner rather than later.

Notes

1 The former Australian prime minister, Kevin Rudd, and opposition leader, Malcolm Turnbull, were both unseated by their parties over disputes on a national emission trading scheme, the Carbon Pollution Reduction Scheme.
2 Energy-intensive sectors were not fully compensated by the CCL reduction, while sectors with low energy requirements and high employment were over-compensated. This was balanced by an 80 per cent rebate in CCL for energy-intensive firms; CCL rates were also frozen between 2001 and 2007, before becoming indexed to inflation. A reduction in the CCL rebate to 65 per cent planned for 2012 again shows how policy packages can evolve over time.
3 In 2010, the newly elected Conservative-Liberal Coalition government announced that it would replace the Infrastructure Planning Commission (IPC) with a Major Infastructure Planning Unit within the Planning Inspectorate. The new unit is still intended to facilitate fast-track decision-making on major infastructure projects but the Coalition claims, with more democratic process than would have operated under the IPC.

References

Anable, J. and Shaw, J. (2007) Priorities, policies and (time)scales: the delivery of emissions reductions in the UK transport sector. *Area* 39 (4): 443–457.
Bailey, I. (2007) Neorealism, climate governance and the scalar politics of EU emissions trading *Area* 39 (4): 431–442.
Bailey, I. and Rupp, S. (2005) Geography and climate policy: a comparative assessment of 'new' environmental polity instruments in the UK and Germany, *Geoforum* 36 (3): 387–401.
Bailey, I. and Wilson, G. (2009) Theorising transitional pathways in response to climate change: technocentrism, ecocentrism, and the carbon economy. *Environment and Planning A* 41 (10): 2324–2341.
Biermann, F. and Pattberg, P. (2008) Global environmental governance: taking stock and moving forward. *Annual Review of Environment and Resources* 33: 277–294.
Börzel, T. (1998) Organizing Babylon: on the different conceptions of policy networks. *Public Administration* 76 (2): 233–273.
Boyd, E. (2009) Governing the Clean Development Mechanism: global rhetoric versus local realities in carbon sequestration projects. *Environment and Planning A* 41 (10): 2380–2395.
Boykoff, M. (ed.) (2009) *The Politics of Climate Change: A Survey*, London: Routledge.

Bulkeley, H. (2005) Reconfiguring environmental governance: towards a politics of scales and networks. *Political Geography* 24 (8): 875–902.

Bulkeley, H. and Newell, P. (2010) *Governing Climate Change*, London: Routledge.

Castree, N. (2010) Crisis, continuity and change: neoliberalism, the left and the future of capitalism. *Antipode* 42 (1): 1327–1355.

Compston, H. (2009) Networks, resources, political strategy and climate policy. *Environmental Politics* 18 (5): 727–746.

Compston, H. and Bailey, I. (eds) (2008) *Turning Down the Heat: The Politics of Climate Policy in Affluent Democracies*, Basingstoke: Palgrave Macmillan.

DEFRA (2008) *A Framework for Pro-environmental Behaviours: Report*, London: The Stationery Office.

Dowding, K. (2001) There must be end to confusion: policy networks, intellectual fatigue, and the need for political science methods courses in British universities. *Political Studies* 49 (1): 89–105.

Gavin, N. (2009) Addressing climate change: a media perspective. *Environmental Politics* 18 (5): 765–780.

Giddens, A. (2009) *The Politics of Climate Change*, Cambridge: Polity Press.

Giddens, A., Latham, S. and Liddle, R. (eds) (2009) *Building a Low-carbon Future: The Politics of Climate Change*. London: Policy Network.

Gouldson, A. and Bebbington, J. (2007) Corporations and the governance of environmental risks. *Environment and Planning C* 25 (1): 4–20.

Grant, W., Matthews, D. and Newell, P. (2000) *The Effectiveness of European Union Environmental Policy*, London: Macmillan.

Hajer, M. (1995) *The Politics of Environmental Discourse*, Oxford: Oxford University Press.

Hale, S. (2010) The new politics of climate change: why we are failing and how we will succeed. *Environmental Politics* 19 (2): 255–275.

Intergovernmental Panel on Climate Change (IPCC) (2007) *Climate Change 2007: Mitigation of Climate Change*, Cambridge: Cambridge University Press.

Jordan, A., Huitema, D., van Asselt, H., Rayner, T. and Berkhout, F. (eds) (2010) *Climate Change Policy in the European Union: Confronting the Dilemmas of Mitigation and Adaptation?*, Cambridge: Cambridge University Press.

Kenis, P. and Raab, J. (2003) Wanted: a good network theory of policy making. Paper to the 7th National Public Management Conference, Washington, DC, 9–10 October.

Kerr, A. (2007) Serendipity is not a strategy: the impact of national climate programmes on greenhouse-gas emissions. *Area* 39 (4): 418–430.

Kickert, W. and Koppejan, J. (1997) Publich management and network management: an overview. In Kickert, W., Klihn, E.-H. and Koppenjan, J. (eds) *Managing Complex NetworksÚ Strategies for the Public Sector*, London: sage (pp. 35–61).

Klijn, E.-H. and Koppenjan, J. (2000) Public management and policy networks. *Public Management* 2 (2): 135–158.

Klijn, E.-H., Koppenjan, J. and Termeer, K. (1995) Managing networks in the public sector: a theoretical study of management strategies in policy networks. *Public Administration* 73 (3): 437–454.

Lohmann, L. (2005) Marketing and making carbon dumps: commodification, calculation and counterfactuals in climate change mitigation. *Science as Culture* 14 (3): 203–235.

Macdonald, D. (2008) Explaining the failure of Canadian climate policy, in Compston, H. and Bailey, I. (eds) *Turniing Down the Heat: The Politics of Climate Policy in Affluent Democracies*, Basingstoke: Macmillan (pp. 223–240).

Newell, P. (2009) Varieties of CDM governance: some reflections. *Journal of Environment and Development* 18 (4): 425–435.

Newell, P. and Paterson, M. (2010) *Climate Capitalism*, Cambridge: Cambridge University Press.

Niemeyer, S., Petts, J., Hobson, K. and McGregor, G. (2004) Understanding thresholds in human behaviour and responses to rapid climate change, *ESRC Environment and Human Behaviour Programme Working Paper* 04/01, University of Birmingham.

Nunan, F. (1999) Barriers to the use of voluntary agreements: a case study of the development of packaging waste regulations in the UK. *European Environment* 9 (6): 238–248.

Pacala, S. and Socolow, R. (2004) Stabilization wedges: solving the climate problem for the next 50 years with current technologies. *Science* 305 (5686): 968–972.

Pattberg, P. and Stripple, J. (2008) Beyond the public and private divide: remapping transnational climate governance in the 21st century. *International Environmental Agreements* 8 (4): 367–388.

Paulsson, E. (2009) A review of the CDM literature: from fine-tuning to critical scrutiny? *International Environmental Agreements: Politics, Law and Economics* 9 (1): 63–80.

Pendleton, A. (2010) After Copenhagen. *Public Policy Research* 16 (4): 210–217.

Pralle, S. (2009) Agenda-setting and climate change. *Environmental Politics* 18 (5): 781–799.

Rhodes, R. (1985) Power-dependence, policy communities and intergovernmental networks. *Public Administration Bulletin* 49 (1): 4–31.

Skjærseth, J. and Wettestad, J. (2008) *EU Emissions Trading: Initiation, Decision-making and Implementation*, Aldershot: Ashgate.

Stern, N. (2007) *The Economics of Climate Change: The Stern Review*, Cambridge: Cambridge University Press.

Stern, N. (2009) *A Blueprint for a Safer Planet*, London: The Bodley Head.

Thompson, A. (2006) Management under anarchy: the international politics of climate change. *Climatic Change* 78 (1): 7–29.

van Waarden, F. (1992) Dimensions and types of policy networks. *European Journal of Political Research* 21 (1): 29–52.

Warner, L. (2007) Einstein, Roosevelt and the atom bomb: lessons learned for scientists communicating climate change, in Moser, S. and Dilling, L. (eds) *Creating a Climate for Change: Communicating Climate Change and Facilitating Social Change*, Cambridge: Cambridge University Press (pp. 167–179).

While, A., Jonas, A. and Gibbs, D. (2010) From sustainable development to carbon control: eco-state restructuring and the politics of urban and regional development. *Transactions of the Institute of British Geographers* 35 (1): 76–93.

Part V
Conclusion

13 Conclusions

Alienation, reclamation and a radical vision

David Manuel-Navarrete, Mark Pelling and Michael Redclift

This book has explored responses to the combined challenge of capitalism's recurring economic crises and contemporary global environmental change from reformist and radical perspectives. The reformist approach is predominant and seeks to keep responses to crises within the bounds of established economic and social relations, with fundamental systems rules not challenged. Radical alternatives see transformation in the structure of governance regimes and forms of agency as necessary components in moving towards a more resilient and sustainable future.

The richer world's responses to climate change and the financial crisis of 2008 have been dominated by reformist discourses in which the hegemony of the current capitalist mode of production provides the core narrative to resilience, and shapes the basis upon which crisis and vulnerability is understood, interpreted and acted upon. Under the reformist worldview the main changes required are technological and administrative, and therefore appear as a matter of efficiency, bureaucratic reorganization and policy-making. This has led policy attention to look towards the accelerated adoption of green technologies and innovation in insurance and engineering solutions to meet the challenge of environmental change, while coping with economic fluctuation or business cycles through either reducing public social spending (to bail out banks) or perhaps, following the insights of Keynesian analysis, expanding public investment in times of market contraction. In fact the combined financial and environmental crisis was consistently presented by mainstream media and governments as a win–win opportunity to stimulate the economy through investment which would simultaneously transform the energy grid with renewables. This win–win strategy required sustained stimulus packages, which in retrospect only some countries, notably the USA and China, were willing or in a position to provide. In most of Europe, where public spending has been severely cut, the financial crisis has restricted investment in and adoption of renewables, and shifted the attention of politicians and the media away from environmental priorities.

At the time of writing, the latest turn in energy policy demonstrates well the unpredictability of policy pathways. The devastating East Japan earthquake, tsunami and associated radiation leak from the Fukushima reactor in 2011 have had global repercussions, and in Europe this is exemplified by Germany's withdrawal from new nuclear investment and increased support for renewables: not an insignificant act for the World's fifth largest economy (World Bank, 2011). Elsewhere, China in particular demonstrates the complexity of energy economics, being both an impressive global-scale investor in renewables and a major net global carbon polluter. Projects such as the Three Gorges dam, which was rejected for funding by the World Bank, and has displaced over one million people, arguably signifies both the scale of China's ambition to respond to the climate change challenge and establish domestic energy security, and the catastrophic impacts such driven responses can have for local populations and ecologies. Emerging trends in the marketization of national food and resource security policy through the buying of long-term land and resource rights hints at new ways in which the balance of security between richer and poorer countries will unfold. In 2011, for example, Bangladesh began a process of acquiring the leases for large areas of African land as part of this flood-prone country's national strategy for food security. Over 600,000 hectares are reported being under discussion (Reuters, 2011). This is an adaptive response with implications for both countries, and builds on a practice of national resource security through control of production spearheaded by Africa, Latin America and elsewhere by China.

Given the dynamism of global economic and associated political relations outlined above, it is perhaps understandable for politicians and citizens alike to search for a degree of stability and to equate this with security. The view that both economic fluctuations and environmental degradation can be kept under reasonably sustainable, or at least manageable limits without radically changing the prevalent institutional order has consequently become popular and captured the articulation of policy under the name of resilience (Brown, Chapter 3, this volume). But is institutional reform sufficient? If not, under which conditions is a radical approach more likely to take place? Returning to Chapter 1, Handmer and Dovers (1996) warn us that established political and socio-economic systems are hard to shift; path-dependency embodied in institutions as well as physical infrastructure and assumptions about the way life should be are considerable challenges to be faced by any agenda that seeks more than a 'tinkering at the margins'. Yet, the scale of threat associated with the combined economic and ecological expressions of crisis point to such deep-rooted concerns with the dominant mode of capitalism. Accordingly, we argue that the task of social science is now more than ever to look beyond the prevalent rules and imagine what an alternative, low-carbon, high-equity and low-risk future might look like – and how we might get there. We need critical theories of transformational social change that go beyond hopes for a technological solution and liberal democracy policy.

A grand transformation is required on a par with humanity's movement from hunter-gather to sedentary agriculture, the enlightenment or the industrial revolution. Taking on the scale of the challenge, and following Marx´s insights on the transition from feudalism to capitalism, David Harvey (2009) advances a 'co-revolutionary theory'. This proposal was noted in Chapter 1 and we turn to it again here as a starting point for a radical transformational view. Harvey's co-revolutionary theory proposes studying the dialectical unfolding of relations between seven moments of social change: (1) technological and organizational forms of production, exchange, and consumption; (2) relations to nature; (3) social relations between people; (4) mental conceptions of the world, embracing knowledges and cultural understandings and beliefs; (5) labour processes and production of specific goods, geographies, services, or affects; (6) institutional, legal and governmental arrangements, and (7) the conduct of daily life that underpins social reproduction. Each one of these moments is internally dynamic and internally marked by tensions and contradictions, but all of them are co-dependent and co-evolve in relation to each other. In the case of political ecology, Harvey's approach can be linked to the Common Pool Resource theories and the Diagnostic Approach proposed by political scientist (and Nobel Laureate in Economics) Elinor Ostrom (2007, 2008) that explained the emergence of capitalist production from communal forms of resource exploitation. Ostrom drew on game theory while Harvey writes a grand theory in political economy.

Social sciences can contribute with theories and straightforward methodologies to understand social change praxis, including its informal, personal and human dimensions. In particular, we argue that research should pay further attention to the role of human agency and the processes of alienation from ourselves, society and nature. Complexity theory shows that 'agents' in chemical, biological and management systems self-organize by following simple interaction rules from which complex patterns emerge (as outlined in Chapter 1). Human agents are heterogeneous (not like the *Homo Oeconomicus*) and enjoy much higher degrees of freedom to choose, and create, rules of interaction than the agents from biophysical systems. However, these crucial moments of choice in which emancipated or alienated individuals and collectivities seize, or not, opportunities for change seem crucial for the study of grand social transitions. This chapter synthesizes key points made throughout the book and organizes these against existing proposals for reform and more radical responses that begin to point to how people in rich and poorer societies might approach the grand transformation that must lie not too far ahead.

From the dead-end of reform to an emerging radical critique

Climate change and development policies have consistently responded to global crises through the promotion of mitigation and compensation policies and incremental reforms that aim at increasing resilience of contemporary socio-political structures, such as minor increases in the capital requirements of banks. As discussed by Carson (Chapter 5, this volume) in the case of the USA, transformation under Liberal Democracy is made difficult because of the strength of interests invested in the status quo; this is so even for apparently basic policy reforms such as the adoption of carbon-free technologies. In fact, systemic shifts in policy making are not necessarily dependent on the outcome of elections or amenable to mass/popular will. Consequently, Bailey and Compston (Chapter

12, this volume), perhaps echoing Giddens (2009) and others (see Chapter 1), argue that new forms of climate politics need to be created before we have any chance of seeing institutional reform that effectively manages the tensions between the demands from consumer capitalism and global environmental change. It is suggested that the emergence of this new climate politics would benefit from reflection (including systematic research) that places attention in the day-to-day practicalities involved in reforming institutional structures and process as a means to unblock political obstacles to stronger climate policy. In a similar direction, Gouldson and Sullivan (Chapter 8, this volume) discuss the need to enlarge 'spaces for feasible action' on climate change as a way of accelerating the process of ecological modernization. These authors still adopt a reformist approach to power (Manuel-Navarrete, 2010): they explore feasible action within the context of existing institutions and power structures, and assume continued economic growth as the goal for development. Thus, current socio-political structures and forms of agency are taken as a given as the analysis can then focus on practical and feasible actions, especially those involving technological and politically neutral organizational innovations.

Other contributors to this book argue that addressing the causes of the financial and ecological crises needs a more critical and fundamental approach to the analysis of current institutional frameworks. Redclift (Chapter 2, this volume) suggests that the global financial architecture, which requires ever-increasing global consumption and achieves it through unsustainable credit mechanisms, is now at stake. Together with North and Scott Cato (Chapter 7, this volume), Redclift responds to Giddens' (2009) call for a new politics of climate change and Gouldson and Sullivan's emphasis on the need to open new spaces for feasible action by exploring the structural changes that may emerge from the bottom-up, through social and political mobilization and community-based strategies, as people push politicians to significantly advance decarbonization agendas. In a similar vein, Mitlin explores lessons learned from collective action experiences in large urban centres in Asia that show how through collective social action society and underling power systems can be reorganizing in ways that distribute influence to the organized poor and enable more equitable and environmentally sustainable (and low-risk) development choices to be made. Here current modes of Asian capitalism are not destabilized but new political spaces are opened up for local alternatives to emerge. There is resistance but this has been overcome by strong and committed leaderships and mass local popular support in conditions of extreme inequality. A key challenge identified by North and Scott Cato, Mitlin and Redclift is to break through current materialist and individualist structures that prevent alternative – collaborative, ecologically sound and egalitarian – modes of development from flourishing. The Transition Initiatives 'movement', analysed by North and Scott Cato, might be already paving this path through generating visions, ideas and techniques for living in utopian post-carbon communities. This is less clear in Mitlin's case where development visions remain concerned with increasing material quality of life, though through collective as well as individual models of consumption.

Utopian thought clearly adopts a radical approach to power in contrast to the reformist stance of ecological modernization (Manuel-Navarrete, 2010). Hjerpe and Linnér (Chapter 11, this volume) consider utopian thought as key for unleashing societal creativity and appreciating the future in its potentially rich diversity, including leapfrog changes in lifestyles, production and consumption. In their view, mainstream exercises to think about the future tend to end up too constrained by assumptions and methodology. Proper utopias should not only envisage the external features of low-carbon functional societies, but also refer to the power relations and dominating values that might achieve and maintain such change. In square opposition to the realism of ecological modernization, Barry (Chapter 9, this volume) positions structural changes of presently asymmetrical power and knowledge relations at the centre of his discussion about moving beyond growth and transcending carbon-fuelled capitalism. He proposes a new political economy based on the notion of 'economic security' as an attractive and radical way of challenging the hegemony of economic growth and of addressing climate change and global inequality. Manuel-Navarrete (Chapter 10, this volume) locates the origin of the combined ecological and economic crises in the dependence on growth and the forms of alienation from ourselves, society and nature, which are necessarily associated with capitalist forms of organization. For these authors, the causes of global environmental change and social inequality are structurally coupled with current capital accumulation patterns, which are fostered by current forms of democracy, and whose fluctuations are expected to become increasingly recurrent and unmanageable.

Far from seeking to claim the superiority of certain forms of analysis over others, this book shows a wide range of available insights and proposals. Reformist approaches may help in directing government and corporation's actions towards, for instance, regulating markets, creating economic incentives, changing consumer behaviours, improving governance or fostering social learning in order to accelerate the rate of adoption of carbon-free technologies and preventing economic fluctuations from slowing down that rate of adoption. These are feasible actions that can be implemented independently of the eventual materialization of positive and radical structural changes. However, the very materialization of these structural changes may well depend on critical analyses that offset the current methodological hegemony of economics and liberal politics.

The critical position might start from the supposition that government's role to defend the public good over private interests ignores the bourgeoisie origin of the state and the historical influence of political economy in its formation and development. This makes unlikely any leadership role to be expected of the state in meeting the economic-ecological crisis that would go beyond tinkering. Yet, in order to effectively transcend, a more nuanced analysis of power and social change, including the interplay between agency and structure, will be needed. This will need to include an understanding of the way in which government institutions function and how they might be influenced from within as well as how far political discourse can effect strategic policy and material change, and how and who is well placed to seize the initiative in making meaningful (and not only symbolic) bottom-up transformations that can lead shifts in development. Some progress is being made in this direction by global environmental change research that has applied understandings of resilience based in socio-ecological systems theory that incorporate an analysis of power and agency (Pelling and Manuel-Navarrete, 2011) (see also Brown, Chapter 3, and Pelling, Chapter 4, this volume). Within modernized societies, consumerist views of human agency greatly limit the scope for individuals and collective critical self-reflection as a starting point for social change. Freire's (1969) distinction between 'adapted man' (*sic*), who works at better aligning himself with the prevalent system, and critical consciousness, which arises once one's position within prevalent orders is realized (opening up the possibility for action to resist or transform the constraining structures of life), is never more apt. In response to consumerism's alienating and individualizing consequences analysts have looked beyond individuals to locate agency and leadership in transformation within social movements and the autonomous organization of local communities (Martínez-Alier *et al.*, 2010). This leads to an analysis of the socio-political conditions which determine active citizenship and increase the influence of grassroots and social-economy organizations over the (local) economy.

Alienation – and reclamation

This book calls for a reclaiming of self, society and nature. This is in response to the above analysis and in the preceding chapters which indicate to us that at root the combined crisis is a feature of the alienating effects of contemporary capitalism – alienation that may be seen to act on the self, to separate individuals from one another and from the natural world. Alienation may be seen to unfold in different ways – and these can be reinforcing (even legitimating and certainly normalizing of alienation). Marx warned that capitalism leads to the alienation from our own inner development, that of others and the environment (Marx, [1844] 1959). This was one of the main concerns of the so-called 'young Marx', rediscovered in the 1960s by Fromm (1961) and others, and later reformulated in terms of political ecology (Gorz, 1980). In this process, it is in repeated acts of production that we subordinate our own creative capacities to the things that we have ourselves created:

> Today entities like money, capital, and the state are crazily accepted as subjects; at the same time, we treat each other and ourselves, not as free self-creating subjects, but as if we were things. That is how we necessarily cut ourselves off from understanding ourselves.
>
> (Smith, 2005: 159)

Thus, subjugation to material growth does a disservice to authentic freedom. To make things worse, it is becoming increasingly clear that due to the finiteness of the planet, growth may only provide local security for some and is in fact raising

global insecurity. Given our increasing power to mould not only social structures, but the very foundations of ecological organization, the discussion of ways
to supersede alienation by reconnecting with ourselves, society and nature
becomes crucial for a meaningful formulation of global economic and environmental politics.

Like Marx, Jacques Ellul (1988) described alienation as a process of individualization and atomization. Ellul uses the notion of *technique* to describe the ways
in which administrative and technological apparatus are inserted between
humanity and the surrounding life-world. Revisiting Ellul for a treatise on
degrowth, Martínez Alier *et al.* (2010) describe technique as producing alienation effects as modern man (*sic*) becomes an 'instrument of his own instruments'. The proposed solution to this form of alienation is to promote quality of
life and collective identity over ideologies of productivity and individuality.
Jackson (2009) agrees that a shift from growth/productivity to a more rounded
notion of 'prosperity', close to Ellul's quality of life, is desirable to help shift
society away from the black hole of growth addiction. But much here depends
on the content of the alternative norms and behaviours to be fostered and how
they are maintained. There is agreement on the reinforcing qualities of a more
rounded philosophy of prosperity and that this would include and be sustained
by stronger collective and community ties. How to achieve a social critical mass
and how far this may be expected to arise from within contemporary capitalist
forms is less clear. These are critical questions for our times however.

Alienation is not only produced by production systems; it can also arise from
consumption. Consumption has become a powerful motor for the economy, and
at the same time has come to provide the dominant currency for expressing individual identity, a language for social interaction and, through market 'choices'
for organic, locally sourced or fair-trade produce, consumption has come to
mediate relationships with nature. However, and especially in more affluent societies, these potential advantages of consumption (e.g. to provide efficient and
relatively easily accessed articulations of an individual's values) have become
more important than the values themselves. What we express through consumption is tightly bounded by what material consumption can offer. This is a key
critique of the ethical consumption paradigm – choices may be individually progressive within a consumer-led economy but they do not necessarily point to a
future that is less materially dependent.

Alienation by consumption is a problem because it makes any shift away
from the current model of a resource- (and debt-) intensive economy more difficult. Undoing commodity fetishism may not be easy but looks to be an important
component of a shift in living that can reduce exposure to economic and ecological crises. What might this transformation look like? Some of the work in this
volume – for example, on Transition Towns – presents ways in which committed communities have already begun to organize less materially dependent
living, and where community offers a more direct way of expressing self and
engaging society than is true for the mainstream. But such experiments are
bound by the market and regulatory as well as the wider cultural context. They

are important signifiers of public concern and experiments in accessible, altern-ative living but the scale of economic and ecological threat we face requires larger solutions, ones that take on a truly democratic state and can enable it to regain some leadership and control of market drivers. Other individual, market-led solutions, based, for example, on ethical consumption, are also important but limited for the reasons expressed above. But what might transitions to state-led or -supported transformation look like? How might transformation in behaviour unfold between scales (with transformative change at one level potentially allowing resilience at another so that transition may be unblocked – or applied in an evolutionary fashion to explore unforeseen consequences)? What would the geography of this look like on the ground – can different economic and social forms with their attendant material expressions co-exist?

One response to the question of strategic leadership has been offered by the German Advisory Council on Global Change (WBGU, 2011) in proposing a new Social Contract for Sustainability. The WBGU argue for 'proactive states with extended participation opportunities'. This formulation brings together an empowered state and an active citizenry: two dynamics that under neoliberal constructions of capitalism and democracy have come to be assumed to be in opposition. Essentially this formulation is a call for strengthened states with greater power to regulate and protect public goods and services while at the same time broadening citizen participation to avoid strength leading to centralization of power. This is one way in which social alienation may be confronted but relies upon shared values and an aspiration for a radical version of sustainability. WBGU (2011) are detailed in their proposals for how such a transition in values, institutional structures and development pathway might unfold. They present a novel vision of responsibilities between the private, state and civil sectors with active and communication flow, cooperation and the dissemination of innovation as the fuel for change. Importantly this vision targets transition in the interna-tional community and at the regional level (in this case the European Union) to set the institutional context for national transformations. This said, the Social Contract for Sustainability proposal remains largely reformist in approach pro-moting carbon markets, renewables, etc. Less is said about the role of local rela-tions between the grassroots and the state in generating popular support for an intellectual capital that might power such a transition – the view so far is top-down, though at the time of writing the WBGU had not yet reported on its full vision.

The necessity for the state to adopt a leadership role to meet the scale of the challenges we face has become a theme in work considering transitions and transformation. The importance of reinvigorating public life and of providing new, meaningful and engaging ways for people to interact and so build a greater sense of collective endeavour is central to the work of Tim Jackson (2009) in his call for a paradigm shift towards a way of living that provides for prosperity without economic growth. Writing from a UK perspective, Jackson champions an agenda for revitalizing existing public goods (green spaces, parks, youth centres, libraries, museums, festivals, local markets, etc.) as a mechanism to catalyse a renaissance in social participation and to cultivate common citizenship. Extending this argument is a call for active citizenship expressed through commitment and participation in public healthcare, education and transport. But interestingly this is not a simple case of the pendulum swinging between calls for private and public political-development models. Enhancing the quality and sense of ownership invested in public goods is an important option for counteracting the social preference for private affluence, and all the consumption and resource consequences, to say nothing of the social inequalities, so generated.

Conclusion: an impasse and beyond

These are pressing times, but exciting also, with the arguments made in this book pointing towards a deep reflection to be made on the balance between public and private lives as well as on quality and quantity in material life. More interesting still are the proposals being made that do not present 'either–or' solutions but rather suggest new hybridities, for example. where greater citizen participation may be seen to strengthen and not weaken the state. Although this may reflect our collective inability to approach utopia and a constrained radical frame, it may as much reflect a strategic vision. Rapid transformations are seldom without significant social and political risk. Our contention is that transformation is inevitable and will either be forced by crisis or may be chosen. But the time window for making this choice is upon us now, and may not be open for too long. Certainly the gradual pace of reforming change appears uncomfortably slow when confronted with the findings of the IPCC – itself a conservative body that has time and again increased its certainty regarding the severity and speed with which dangerous and irreversible climate change is approaching.

Our discussions have pointed to the scope for radical change that can and arguably needs to be made in all aspects of contemporary late modern life. We use alienation as a lens on this and find alienating processes operating at the moments of production, consumption and through the intervention of administrative efficiency and technology in everyday life. The impasse we face is less one of analysis than of motivation. Confronted by limited action and weak commitment from governments, and by societies fragmented through individualization, the search for leadership has moved to the organized civil sectors. Civil society has shown itself capable of organizing local alternative economic and political forms that are more equitable and inclusive, and point to pathways for individuals to overcome their own alienations. Some local actors have explored the possibility of shaping the way in which bureaucracy and technology is used to provide goods and services, stimulating local food and energy production while coupling this with cultural identity struggles and alternative consumption systems. But these need to be scaled up to meet the challenge we face, and this is difficult without state involvement and new forms of political agency. Radical vision and brave leadership is needed to move past this impasse.

References

Ellul, J. (1988) *The Technological Bluff*, Grand Rapids, CO: Eerdmans.

Freire, P. ([1969] 2000) *Education for Critical Consciousness*, New York: Continuum.

Fromm, E. (1961) *Marx's Concept of Man*, New York: Frederick Ungar.

German Advisory Council on Global Change (WBGU) (2011) *World in Transition: A Social Contract for Sustainability*, Summary for Policy Makers, accessed from www.wbgu.de/en/publications/flagship-reports/flagship-report-2011/.

Giddens, A. (1984) *The Construction of Society*, Cambridge: Polity Press.

Giddens, A. (2009) *The Politics of Climate Change*, Cambridge: Polity Press.

Gorz, A. (1980) *Ecology as Politics*, London: Pluto Press.

Handmer, J.W. and Dovers, S.R. (1996) A typology of resilience: rethinking institutions for sustainable development. *Organization and Environment* 9 (4): 482–511.

Harvey, D. (2009) *A Companion to Marx's Capital*, New York: Verso.

Jackson, T. (2009) *Prosperity Without Growth: Economics for a Finite Planet*, London: Earthscan.

Manuel-Navarrete, D. (2010) Power, realism, and the humanist ideal of emancipation in a climate of change. *Wiley Interdisciplinary Reviews: Climate Change* 1: 781–785.

Martinez Alier, J., Pascual, U., Vivien, F.-D. and Zaccai, E. (2010) Sustainable degrowth: Mapping the context, criticism and future prospects of an emergent paradigm. *Ecological Economics* 69: 1741–1747.

Marx, K. ([1844] 1959) *Economic and Philosophic Manuscripts*, Moscow: Progress Publishers. First Manuscript.

Ostrom, E. (2007) A diagnostic approach for going beyond panaceas. *PNAS* 104 (39): 15181–15187.

Ostrom, E. (2008) The challenge of common-pool resources. *Environment: Science and Policy for Sustainable Development* 50 (4): 8–21.

Pelling, M. and Manuel-Navarrete, D. (2011) From resilience to transformation: exploring the adaptive cycle in two Mexican urban centres. *Ecology and Society* 16 (2): 11. Available at www.ecologyandsociety.org/vol. 16/iss2/art11.

Reuters (2011) Bangladesh looks to Africa to boost food output. Accessed online at http://in.reuters.com/article/2011/05/19/idINIndia-57132020110519.

Smith, C. (2005) *Karl Marx and the Future of the Human*, Lanham, MD: Lexington Books.

World Bank (2011) indicators database. Accessed online at http://siteresources.worldbank.org/DATASTATISTICS/Resources/GDP_PPP.pdf.

Index

Page numbers in *italics* denote tables, those in **bold** denote figures.

Taylor & Francis

eBooks

F O R   L I B R A R I E S

ORDER YOUR FREE 30 DAY INSTITUTIONAL TRIAL TODAY!

Over 23,000 eBook titles in the Humanities, Social Sciences, STM and Law from some of the world's leading imprints.

Choose from a range of subject packages or create your own!

Benefits for **you**

- ▶ Free MARC records
- ▶ COUNTER-compliant usage statistics
- ▶ Flexible purchase and pricing options

Benefits for your **user**

- ▶ Off-site, anytime access via Athens or referring URL
- ▶ Print or copy pages or chapters
- ▶ Full content search
- ▶ Bookmark, highlight and annotate text
- ▶ Access to thousands of pages of quality research at the click of a button

For more information, pricing enquiries or to order a free trial, contact your local online sales team.

UK and Rest of World: **online.sales@tandf.co.uk**

US, Canada and Latin America:
e-reference@taylorandfrancis.com

www.ebooksubscriptions.com

ALPSP Award for BEST eBOOK PUBLISHER 2009 Finalist

Taylor & Francis **eBooks**
Taylor & Francis Group

A flexible and dynamic resource for teaching, learning and research.